Embodied Carbon

Embodied Carbon

Carbon Justice in the Built Environment

MING HU

OXFORD
UNIVERSITY PRESS

OXFORD
UNIVERSITY PRESS

Great Clarendon Street, Oxford, OX2 6DP,
United Kingdom

Oxford University Press is a department of the University of Oxford.
It furthers the University's objective of excellence in research, scholarship,
and education by publishing worldwide. Oxford is a registered trade mark of
Oxford University Press in the UK and in certain other countries

Published in the United States of America by Oxford University Press
198 Madison Avenue, New York, NY 10016, United States of America

British Library Cataloguing in Publication Data
Data available

Library of Congress Control Number: 2024950792

ISBN 9780198944454

DOI: 10.1093/9780198944485.001.0001

Printed and bound by
CPI Group (UK) Ltd, Croydon, CR0 4YY

Contents

Foreword

Buildings are central to a carbon neutral future. In the US, over 70% of electricity production is consumed by equipment and activities in buildings. Since the power plants that produce the electricity are typically only 30% efficient, every kilowatt-hour saved in a building is three times more valuable at the powerplant. Buildings also unnecessarily consume over 35% of the nation's fossil fuel production, fuel used for low-grade energy demands such as heating and hot water. Every action taken to reduce energy demand in the building sector is a major contribution to reducing our carbon footprint. We need to continue to aggressively pursue energy efficiency in existing and new construction—down to levels that can be met with local renewables by 2050—the definition of carbon neutrality. A recent report by the US National Academy of Sciences, *Accelerating Decarbonization in the United States: Technology, Policy, and Societal Dimensions*, outlines critical actions needed to accelerate building energy efficiency, electrification and decarbonized energy sources. Buildings must be central to the US commitment to carbon neutrality, indeed to all country's commitments.

Beyond the expansive operational energy use in buildings, the undervalued demands of embodied energy, and the commensurate carbon consequences, crucially needed study. The delayed emergence of debate about embodied carbon is the result of the overwhelming carbon totals generated by operational energy demands in our existing and even newer building stock, only marginally reduced with our slow adoption of stronger standards. However, cutting-edge architects and engineers have created carbon-neutral buildings with energy demands reduced to such low levels that on-site renewables can meet the load, creating truly resilient places to live, work, and study. As new zero energy and zero carbon building codes and standards emerge in a few states and cities, these leading practitioners have turned their focus to the unknown embodied carbon—the carbon generated in the raw material extraction, manufacturing, transport, and demolition of building materials and systems. The Carbon Leadership Forum, the ILFI Zero Carbon standards, BuildingGreen and NBI's Net Zero initiatives, the US and World Green Building Councils, alongside academics and practitioners around the world, are defining approaches to reduce the embodied carbon footprint of the most egregious building materials and assemblies.

This book is the essential policy and practice companion to these global accounting approaches to reduce embodied energy and carbon. In this fourth seminal book, Dr. Ming Hu details that while operational carbon has dominated advances in the building sector for the past five decades, the challenge for today and tomorrow is the combination of operational and embodied carbon, including all the embodied carbon that is 'off-shored' to keep it off the books. Moreover, both operational and embodied carbon have serious environmental and social inequities, material resource depletion challenges, toxic emission challenges, and resiliency challenges by overstating the high tech. Instead, she defines the vital need to re-embrace low-tech, climate-specific, and regenerative solutions to both embodied and operational energy, for a carbon-neutral built environment.

This is a brilliant book on embodied energy and carbon that is a must read for policy leaders, economists, architects and engineers, faculty and students, to fully understand the central role that the built environment plays in the world's carbon footprint. Beginning with the history of life-cycle assessment and embodied carbon, Dr. Ming Hu introduces global embodied carbon labels that have driven advances in low-carbon and regional materials and assemblies. In a series of invaluable chapters, Dr. Hu captures the scale of the embodied carbon challenge, the environmental justice impacts of pursuing rapid reductions in embodied carbon, the vital balance of high tech and low tech, and the policies and incentives that can drive embodied carbon savings. Dr. Hu cautions against an overemphasis on high-tech solutions from carbon sequestration to smart systems and AI. She advances the need to embrace efficiency and embodied carbon through the integration of passive, climate, and culturally appropriate solutions—for sustained advances in the built environment for both the Global North and the Global South.

Fully defining the carbon story of the built environment, this book explores the importance of efficiency and passive strategies, innovation and integration, as well as the importance of designing the system, designing for longevity, and designing local, to achieve a truly carbon-neutral future in the built environment. With flowing writing, Professor Hu advances the argument that addressing embodied carbon is essential to addressing global inequality and achieving carbon neutrality. This book, *Embodied* Carbon, is a sweeping, compelling, and critical text for designing a shared and sustainable future in our built environments.

Vivian Loftness, FAIA
University Professor and Paul Mellon Chair in Architecture,
Carnegie Mellon University
Board Member and former Board Member: International Living Futures Institute, American Institute of Architects, US Green Building Council.

Preface: Why did I want to write this book?

Twenty years ago, during a routine shopping trip at Walmart, I was struck by the sight of a product labelled 'Made in China' with an astonishingly low price tag. This disparity between its cost and the extensive journey it had undertaken—from labour-intensive production to transoceanic shipment—prompted a profound reflection on the true cost and value of consumer goods. How could such a product be so affordable, even cheaper than similar items purchased in China? This incident sparked an enduring unease about the hidden costs of my consumption habits and the broader implications of global trade.

Fourteen years ago, I encountered another startling revelation while working on a super high-rise building in Abu Dhabi with a prominent international architecture and engineering firm. Over 90% of the construction materials were sourced from disparate corners of the globe, including China, Germany, and the United States, with labour primarily imported from India. Despite these extensive logistical efforts, the construction costs remained surprisingly low. This phenomenon led me to question whether globalization inherently entails the exploitation of resources through a skewed economic trading system. Despite its substantial material footprint, the project attained a LEED Gold certification. This paradox made me ponder the genuine cost and value of building, particularly in the context of sustainable architecture.

Around that same time, my mentors at the University of Notre Dame introduced me to the concept of embodied carbon—the carbon emissions associated with the entire life cycle of materials and processes used in building and infrastructure projects, from raw material extraction to demolition, excluding operational emissions. This notion intuitively suggested that buildings constructed with locally sourced materials aligned more closely with sustainable principles due to their lower embodied carbon. It became apparent that traditional, vernacular, and indigenous architectural practices might harbour invaluable insights into sustainable building methodologies characterized by low embodied carbon.

Since then, my scholarly pursuits have centred on the study of embodied carbon—a pivotal concept encompassing the total greenhouse gas (GHG) emissions associated with materials and constructions within the entire life cycle of a building. This includes emissions from material extraction, production, transportation, installation, and the construction process itself. Embodied carbon represents a comprehensive metric for assessing the environmental impact and carbon footprint of the building and construction industry. Notably, embodied carbon can account for up to 50% or more of a building's total carbon emissions, underscoring its critical importance in the sustainability discourse.

The construction industry has made significant strides in addressing operational energy, with well-established carbon emission reduction strategies. However, the domain of embodied carbon remains relatively underexplored. Neglecting this aspect has far-reaching consequences, exacerbating environmental injustices and disproportionately affecting marginalized communities. For instance, consider a scenario where raw materials are extracted in Africa, manufactured into products in China, and ultimately sold in the United States. The carbon emissions embedded in this global supply chain carry profound environmental and social implications. By focusing on embodied carbon, we can champion sustainability, resource efficiency, and environmental responsibility.

Before the advent of the COVID-19 pandemic, I believed that a thorough understanding of embodied carbon merely required the identification of reliable data sources and the refinement of assessment models, perhaps with the aid of artificial intelligence. As an architecture researcher, building scientist, and environmental engineer, my colleagues and I concentrated on scientific evidence and technical solutions. Whilst our approach was not misguided, it was incomplete.

To fully grasp the value, cost, and impact of embodied carbon, it is imperative to consider social, cultural, and political perspectives. The biases of high-tech solutions and the Global North are integral to understanding the complexities of this issue. Consequently, addressing embodied carbon is vital for cultivating sustainable practices, promoting equity, and fortifying ecological resilience. This book delves into the significance of embodied carbon, its distinctive attributes, and the urgent necessity of addressing it to meet carbon reduction targets. It is a collaborative journey of learning and discovering solutions.

This book articulates a compelling argument for a paradigm shift in addressing embodied carbon emissions within the construction sector. It is structured into three parts, each contributing to its central theme:

Part I: Looking Back and Perception (Chapters 1–3) explores the historical evolution of embodied carbon, highlighting its significance in the context of climate change and its role in climate action through illustrative case studies.

Part II: Unravelling the Complexity and the Hidden Inequities (Chapters 4–6) examines the concealed biases surrounding embodied carbon, unveiling their environmental justice implications and the influence of high-tech solutions.

Part III: Environmental Equality for the Future (Chapters 7–9) investigates innovative, equitable, and sustainable strategies to reduce embodied carbon, emphasizing low-tech alternatives and insights from regional and traditional practices.

The aim of this book is to challenge entrenched norms, rectify biases, and inspire a more balanced and responsible approach to mitigating embodied carbon emissions in the built environment.

PART I

LOOKING BACK AND PERCEPTIONS

Chapter 1
Historical evolution of embodied carbon

1.1 Why embodied carbon?

When we think about climate change, we often picture smokestacks and car exhaust, but did you know that the very buildings we live and work in are silent contributors to greenhouse gas emissions? This hidden impact is what we call embodied carbon, and it encompasses all the emissions released over the entire life cycle of a building—from the extraction of raw materials to construction, maintenance, and eventual demolition [1, 2].

Why should we care? Because the building sector is a significant player in the climate crisis, responsible for a substantial portion of global greenhouse gas emissions. Embodied carbon alone accounts for 11% of these emissions [3]. Even more startling, by 2050, nearly 40% of all carbon emissions in the built environment could come from embodied carbon [4].

Understanding embodied carbon isn't just about numbers—it's about equity and justice. Historically, developed regions like the United States and the European Union have contributed the lion's share of emissions (26 and 23% of all cumulative emissions since the Industrial Revolution, respectively), while nations like India and China (3 and 12%), with much larger populations, have contributed far less [5]. This disparity highlights the unequal burden of climate change, where the Global North's past actions disproportionately impact the Global South.

In this book, we'll explore embodied carbon in-depth, adopting definitions from the United Kingdom Parliament to guide our understanding; refer to Box 1.1 [6]:

Box 1.1: Definitions

- **Embodied carbon**: Embodied carbon emissions are all emissions associated with materials, construction, maintenance, repair, demolition, and disposal of a building.

Embodied Carbon. Ming Hu, Oxford University Press. © Ming Hu (2025). DOI: 10.1093/9780198944485.003.0001

- **Whole life cycle**: The entire life of a building from material sourcing, manufacture, construction, use over a given period, demolition, and disposal, including transport emissions and waste disposal.

But embodied carbon isn't just a technical issue. It's a reflection of our values and lifestyle choices. To truly grasp its importance, we'll trace its history—from early concepts and initial resistance to its growing acceptance and critical role today. Join me as we uncover the story of embodied carbon, revealing how our buildings silently shape our climate future and what we can do to build a more sustainable world. This journey is not just about reducing emissions; it's about rethinking how we design, construct, and inhabit our spaces to create a more equitable and resilient world for all.

1.2 Start from embodied energy

The relationship between embodied energy and embodied carbon is intricate and interconnected. Embodied carbon represents the total carbon dioxide equivalent emissions associated with the embodied energy given a specific energy mix and carbon conversion factors [7, 8]. The concept of embodied energy is crucial in evaluating the environmental impact of various processes, especially in building construction and material production. Embodied energy encompasses the total energy consumed throughout a product's life cycle, including energy for raw material extraction, manufacturing, transportation, and disposal [9].

One key distinction between the two is that the product with the same amount of embodied energy can result in different amounts of embodied carbon, depending on carbon conversion coefficient factors [9]. For example, in the United States, like other countries, electricity is produced by many different fuel sources of energy that have varying carbon coefficients. In 2021, in the United States, natural gas has a carbon emission coefficient factor of 52.91kg carbon per million Btu (mmBtu), while coal is 95.92kg carbon per mmBtu. The embodied energy is often measured in kilowatt-hours, and one million Btu is equal to 290.071kWh [10]. The type and amount of emissions produced from one kilowatt-hour of electricity depend on how electricity is generated in different regions. As depicted in Figure 1.1, based on the 2022 data from the Emissions & Generation Resource Integrated Database released in January 2024, within the United States, in the New York City/Westchester (NYCW) subregion, natural gas constitutes 90.1% of the fuel mix. In the SERC

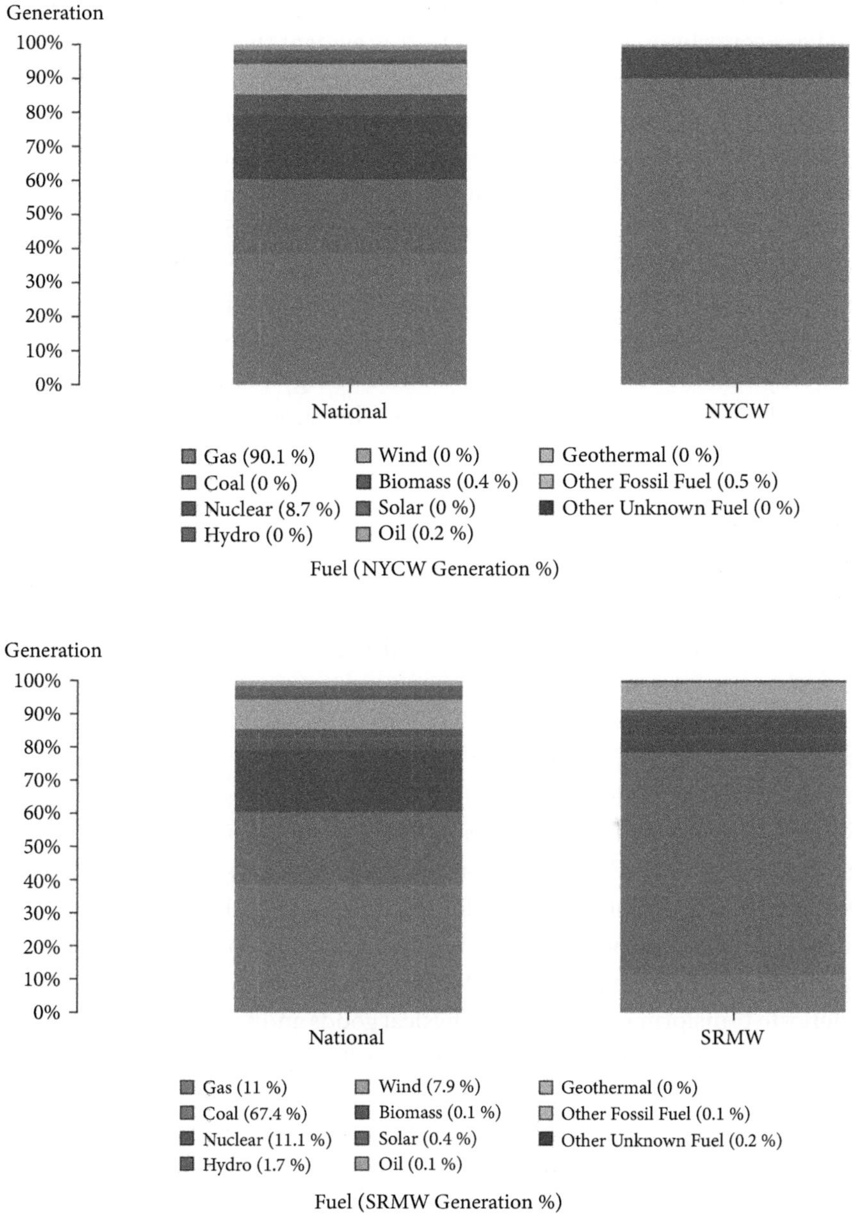

Figure 1.1 eGrid subregions. (left) New York City/Westchester (NYCW) subregion as compared to the national energy generation source mix. (Right) SERC Midwest (SRMW) subregion as compared to the national energy generation source mix.

Reproduced from United States Environmental Protection Agency, Power Profile. (2024). *How clean is the electricity you use?* https://www.epa.gov/egrid/power-profiler#/NYCW

Midwest (SRMW) subregion, coal represents 67.4% of the fuel mix, with natural gas contributing merely 11% [11]. It is evident that generating the same amount of electricity—1 kWh—in the Midwest results in considerably higher carbon emissions than that in New York City. The difference is substantial, with the Midwest's carbon emissions being 1.89 times greater, attributable to differing energy source and carbon conversion coefficients [11].

This variation underscores the criticality of accounting for both embodied energy and embodied carbon when evaluating the environmental footprint of products, processes, and systems. A foundational understanding of embodied carbon begins with an exploration of embodied energy. This is not only because the concept of embodied energy predates that of embodied carbon by several decades but also because, even to this day, the methodologies for assessing, considering, and managing embodied carbon remain inextricably linked with those of embodied energy, for better or for worse.

1.3 The intellectual origins of embodied energy: the pioneers

In 1886, Ludwig Boltzmann (Austrian physicist and philosopher) posited a concept that situated the struggle for available energy at the heart of life's endeavours [12]. Following in his intellectual footsteps, Frederick Soddy, a Scottish scientist and Nobel laureate in chemistry, articulated in 1933 that energy availability is the cornerstone upon which life is sustained and all necessary materials are produced. Soddy put forth one of the earliest arguments linking physics, ecology, and economics, emphasizing the interconnectedness of those fields [13]. Soddy argued that real wealth was derived from the use of energy to transform materials into physical goods and services. Therefore, real wealth formation is distinct from virtual wealth formation—an accumulation process. He also advocated creating a detailed energy accounting structure to track the actual cost of goods and products rather than using the monetary value [14]. This is the predecessor of embodied energy counting.

Even though Soddy's insights laid the foundation for understanding the importance of considering energy flow and sustainable development principles in economic decision-making processes, his theory of energy did not create a substantial impact for the following decades for several reasons. One key factor was the complexity and novelty of his ideas at the time. The concepts of energy cost theory for economic valuation were ahead of their time and not easily understood or accepted by the mainstream economic and scientific communities [15].

Additionally, the interdisciplinary nature of Soddy's work, which bridged chemistry, economics, and environmental science, may have made it challenging for experts in each field to grasp the implications of his theories fully [16]. Lastly, the broader sociopolitical context of the time, including the lack of emphasis on sustainability and environmental concerns, might have hindered the widespread acceptance and adoption of Soddy's energy theories. The prevailing neoclassical economic paradigms focused more on growth and expansion rather than on the sustainable use of resources, which could have marginalized Soddy's ideas on energy limitations and conservation [17].

While direct written evidence may not explicitly demonstrate that Nicholas Georgescu-Roegen was inspired by Frederick Soddy's energy theory, the intellectual linkage between their works, particularly within the realm of energy and economic analysis, is compelling. The seamless integration of thermodynamic principles into economic discourse is a hallmark of both Soddy's pioneering efforts and Georgescu-Roegen's subsequent explorations, underscoring the intrinsic synergy between energetic and economic systems.

Nicholas Georgescu-Roegen, a distinguished Romanian-American economist, mathematician, and statistician, made seminal contributions to the field of economics, most notably through his development of bioeconomics. His journey began in the 1930s with a critical review of the hedonistic valuation framework's shortcomings in neoclassical economics. By the 1960s, he explored the tensions between social and hedonistic valuation, and during the 1970s and 1980s, he expanded his focus to include the conflicts among individual, social, and environmental values [18]. His bioeconomic model foregrounded the necessity of integrating ecological principles, specifically the thermodynamic law of entropy, into economic theory [19].

In the wake of the Communist revolution in 1948, Georgescu-Roegen was compelled to leave his homeland of Romania, subsequently taking up residence in the United States. At Vanderbilt University [20], he joined the economics faculty, and his seminal contributions there significantly propelled the evolution of ecological economics—a discipline that partly germinated from his bioeconomic insights. His analysis underscored the entropic nature of economic transactions and the profound interdependency of economic and ecological systems, thus pioneering a more integrative and sustainable economic methodology [21].

One of his doctoral protégés, Herman Daly, became a prominent figure in the ecological economics arena, profoundly influencing the dialogue on sustainable development [22]. Daly advanced three fundamental prerequisites for sustainability: a *sustainable scale* relative to biocapacity, *equitable distribution*, and *efficient resource allocation* [23]. These principles are now acknowledged as essential considerations for incorporating sustainability into decision-making processes [24].

Daly's propositions, while pivotal, encountered a degree of marginalization and scepticism from his peers in the mainstream economic community, including the experts from the World Bank, which was predominantly focused on growth models. Yet, despite this, Daly's intellectual legacy has laid the groundwork for extensive interdisciplinary research. As Daniel Neill notes, this includes the exploration of social metabolism, environmentally extended input–output analysis, post-growth studies, and the concept of strong sustainability [25]. Among Daly's intellectual contributions, the adoption and extended input–output analysis application in environmental science eventually became a cornerstone for life-cycle assessment methodologies, pivotal for quantifying embodied energy and carbon emissions. Daly's knowledge of input–output analysis can probably traced back to his Ph.D. advisor, Nicholas Georgescu-Roegen.

But the official creator of input–output analysis is Wassily Leontief, a prominent economist and Nobel laureate. The heart of the theoretical model behind input–output analysis is to start from the interindustrial relationships embodied by the flows of goods and services between different parts of the economy. Leontief's work was instrumental in providing a framework for understanding the flow of goods and services in an economy, while Georgescu-Roegen's contributions highlighted the importance of ecological constraints and the finite nature of resource [26]. Leontief and Georgescu-Roegen shared a common interest in challenging traditional economic paradigms. Leontief's empirical approach and emphasis on understanding economic inter-relations resonated with Georgescu-Roegen's critique of mainstream economic theory and his call for a more holistic and sustainable economic framework [27]. Nicholas Georgescu-Roegen is a close colleague and strong supporter of Leontief's work [28]. It is not surprising that Daly could have taken up knowledge from there. All those pioneers' work from Soddy to Daly paved the foundation and set the stage where the embodied energy and carbon research could take place in the 1970s.

1.4 Evolution of embodied energy in the United States

1.4.1 The beginning: energy crisis

In the United States, the term 'embodied carbon' found its roots in 1973, beginning with Dr. Robert A. Herendeen's seminal report, 'The Energy

Cost of Goods and Services', which introduced a model to quantify the energy embedded within consumer goods and services [29]. Funded by the National Science Foundation and developed during Herendeen's tenure at the Oak Ridge National Laboratory, then the University of Illinois' Center for Advanced Computation, this pivotal work leveraged the 1963 Input–Output financial data table from the U.S. Department of Commerce to establish embodied energy coefficients across 362 sectors. Herendeen meticulously clarified: 'Total energy cost denotes all energy inputs in the manufacture/sales chain, including extraction of raw materials, primary processing, transportation, intermediate and final fabrication, and final marketing activities' [29]. His insights laid the groundwork for future exploration into the environmental impacts of the built environment, particularly regarding embodied carbon.

In 1975, Herendeen, along with his colleague Dr. Clark Bullard from the University of Illinois' Center for Advanced Computation, continued to refine this model, extending the assessment of total energy to a broad spectrum of the U.S. economy. This partnership led to significant strides in the field, including Jerry Tanaka's collaborative work in 1976, which applied these methods to evaluate the energy costs of household living in the U.S. [30]. A year later, Herendeen broadened the scope further to the energy costs of building construction, presenting his findings in the *Journal of Energy and Buildings*. This sequence of research, buttressed by National Science Foundation funding, propelled the University of Illinois' Center for Advanced Computation to the forefront of embodied energy research [31].

It was in this intellectually fertile environment that the United States' first handbook for calculating embodied energy was born in the same centre. Herendeen's colleague Clark Bullard spearheaded the creation of the 'Net Energy Analysis: Handbook for Combining Process and Input-Output Analysis' [32], a 1976 publication. Funded by the Energy Research and Development Administration (ERDA), the handbook laid out methodologies for tallying both the direct and indirect energy inputs required to produce goods and services. It underscored the significance of indirect energy, resonating with Herendeen and Tanaka's earlier work, which posited that indirect energy demands often surpass direct energy consumption for the average consumer. The researchers defined that indirect energy 'is consumed indirectly as energy used elsewhere in the economy to produce the other goods and services purchased by consumers' [32]. This is the exact definition of embodied energy, later used as a common term. The handbook emphasized the importance of indirect energy by citing Herendeen and Tanaka's 1975 publication:

> Indirect energy is by no means negligible; the average consumer demands more energy indirectly than directly.

It's clear that during the same period, Herendeen and Bullard were innovating at the University of Illinois; their work inspired a collaborative effort in embodied energy research for building construction.

At the same institution, another team, led by Bruce Hannon and including Richard Stein (a New York-based architect), made strides with their 1976 report 'Energy Use for Building Construction', also financially supported by ERDA. This research built upon and employed the same method and model as Herendeen and Bullard's work, confirming the cohesive direction of this pioneering field. In Chapter 9, the author documents an hour-long interview with Dr. Bullard regarding his work in details.

It is apparent that the University of Illinois' Center for Advanced Computation, Energy Group, was a nexus for embodied energy and carbon research in the built environment, drawing upon the resources of both the NSF and ERDA. The concerted efforts of these scholars, particularly in the aftermath of the 1973 Oil Crisis, catalysed the establishment of the U.S. Department of Energy in 1977, consolidating energy-related functions into one federal entity and signifying a watershed in U.S. energy policy and research. The story of ERDA is integral to this narrative. Created in 1975 as part of the Energy Reorganization Act of 1974, ERDA responded to the 1973 Oil Crisis by assuming functions from the Atomic Energy Commission, excluding those delegated to the Nuclear Regulatory Commission [33]. ERDA's remit was vast, encompassing energy research and development, nuclear weapons programmes, and naval reactors. In 1977, ERDA merged with the Federal Energy Administration to form the Department of Energy (DOE), thereby uniting federal energy initiatives at a pivotal moment in addressing national energy challenges.

1.4.2 The integral role of research in shaping energy policy and consumer behaviour

During an interview the author had with Dr. Bullard, he shared an interesting story about how he got involved in quantifying embodied energy and carbon. The first time he was involved in this work was to illustrate how different lifestyles affect the demand for energy. They observed differences between rich and poor people in their spending on gasoline. Poor people spend more on gasoline, while rich people tend to buy airline tickets. This led them to investigate consumer choices, such as whether to buy beer in returnable bottles, aluminium cans, or throwaway glass bottles. They examined the energy differences and published the findings. Following their publication, they received numerous inquiries from Wall Street companies and investors. They wanted to know whether they should invest in certain products. For example, meat at the grocery store is now sold in styrofoam trays with plastic wrapping, whereas it used to be sold in cardboard trays. Companies needed to decide between paper or plastic meat trays and understand how the price of oil affects the costs of styrofoam and cardboard.

Every company needed information on these indirect effects through the economy. For instance, manufacturing a car used to consume as much energy as the car would use in two years' worth of gasoline. Companies considered making engines out of aluminium instead of steel for lighter, more energy-efficient cars. However, aluminium is more expensive due to the high electricity required for production. To this extent, there is market demand for such research.

During his interview with the author, Dr Bullard said:

> We aimed to understand the energy winners and losers from a policy standpoint, making our study crucial for energy policy analysis. This is why I joined the Carter Administration and served as Director of the Conservation and Renewable Energy Policy Office at the Department of Energy when it was established.

DOE research continued under his supervision; the team compared the energy embodied in exported goods to that in imported goods, assuming the imports had the same energy requirements as if they were manufactured domestically. The results were crucial to the formation of energy-related policy at that time. Dr Bullard's experience perfectly demonstrates that from the onset, embodied energy and carbon research were deeply intertwined with policy formation and were indispensable for understanding consumer behaviour.

1.4.3 The overlooked importance of embodied energy and carbon

Why hasn't the importance of embodied energy and carbon been widely adopted since the 1970s? The simple answer is that after the energy crisis dissipated, the urgency to understand and reduce energy consumption diminished in the eyes of policymakers. The more difficult answer is that the world has not been ready to accept the idea that growth is not the only solution for improving human life and that a growth-oriented approach in neoclassical economics, without a system boundary, is flawed. The embodied energy theory of value is anchored to the principle of energy flow, i.e. energy analysis. A critical aspect of energy analysis is determining the total (direct and indirect) energy required for the production of goods and services. This total value is termed 'embodied energy value'.

The prerequisite for calculating the total embodied value is establishing the system boundary where embodied carbon exists and where energy flows. Early calculations of embodied energy were based on national input–output financial transaction data in the economy, which Leontif invented. In this context, the system boundary is the physical, geographic, and political limit of a nation. The net input (or value-added) vector includes labour, government services, capital services, energy, and other natural resources (raw materials). In this input–output equation, the corresponding financial categories are employee compensation, indirect business taxes, and property-type income. The sum of these net inputs, in dollar units, is the gross national product (GNP). In dollar units, energy (fossil fuels, nuclear fuels, and renewables) comprises a small component of the GNP. This has led some researchers and policymakers to conclude that energy is a minor component of economic production. This foundational theory is used in neoclassical economic development: since energy is a minor component, the economy—or, more accurately, human activity—is treated as a perpetual machine that can continue growing indefinitely.

What constitutes the root of this pervasive misinterpretation? The error lies in both the net input and the defined system boundary. From a physical standpoint, the Earth—the boundary within which humanity resides—receives only one principal net input: solar energy [12]. There are no interplanetary deliveries of labour, materials, or government capital investments to Earth. Consequently, all entities on Earth can be regarded as direct or indirect products of historical and contemporary solar energy. **This fundamental principle underpins energy theory [34], a theory that treats the built environment and human activities as a subset of the natural ecosystem.** In this system, fossil fuels and bio-based materials represent the embodiment of millions of years of solar energy, while renewable sources like wind and rivers capture solar energy from more recent times. Following this logic, humans, too, are products of millennia of solar-powered evolution and are sustained by agriculture that harnesses both current and ancient stored solar energy.

In the pre-industrial era, when goods and services were primarily produced within national boundaries, and the movement of labour and services across borders was very infrequent, using the nation as a system boundary might have seemed reasonable, even though the inputs were inaccurately accounted for. However, in the post-industrial age, with transnational imports and exports being essential to support human activity, the system boundary has become exceedingly vague, exacerbating the miscalculation of net inputs.

This brings us to a pivotal question: does this flawed system boundary lead to an unfair distribution of the burdens of embodied energy consumption and carbon emissions among nations? Regrettably, the answer is affirmative and clearly described in Box 1.2.

Box 1.2: System boundary extension and uncounted environmental burden

Consider a scenario where raw materials are extracted in Africa, processed into products in China, and then sold in the United States. The embodied energy and subsequent environmental impacts of material extraction in Africa, combined with the energy consumption and carbon emissions from manufacturing and transportation in China, are substantial. Yet the consumers of these products, the actual drivers of such economic activity, remain physically disconnected from the associated energy use, emissions, and environmental degradation.

When the Global North touts its carbon reduction achievements, which often result from relocating energy and carbon-intensive industries to the Global South, the burden is inappropriately shifted to those who did not create the demand. This is more than a supply chain issue—it's a profound challenge of distributive equity and environmental justice.

But who is willing to oppose the notion of growth or to limit their access to more and cheaper imported goods and services? Unsurprisingly, no one is eager to volunteer personal sacrifices for the sake of environmental justice. This unpopular political and economic reality hinders the embrace of the embodied carbon concept and has led to its rejection and exclusion from mainstream policy discussions. As the United Kingdom example given by Azari and Moncaster demonstrate, embodied carbon has long been a thorny issue and, for many years, excluded from policy and regulation [35]. In 2007, the UK Government Department for Communities and Local Government (DCLG) published a policy statement, 'Building a Greener Future: Policy Statement'. Despite the previous comments to the draft policy, DCLG completely ignored the embodied carbon:

> 3.1.2 We do not believe a full consideration of embodied carbon is practical or realistic in the short-to-medium term. Evidence on the lifetime carbon costs of particular technologies is weak, and varies considerably depending on where and how they are manufactured. [35]

1.5 Acceptance, adoption, and development

Despite the ignorance from policymakers and mainstream growth-oriented development, in the eyes of researchers the studies on embodied carbon and its broad impact continue. Following the 1970s, great research was carried out in the field of embodied carbon assessment, as mentioned in Section 1.4.3. This knowledge established a solid intellectual foundation for further development in embodied carbon calculation, management, and mitigation. Especially in the past two decades, we have seen exponential growth in research and knowledge in this area.

1.5.1 The development of embodied carbon research in the built environment

The author conducted a literature review for this book to comprehend the extant knowledge of embodied carbon within the built environment. The Web of Science was selected as the research database for its extensive and well-regarded collection of peer-reviewed publications. This search yielded a total of 372 peer-reviewed journal articles. Figure 1.2 depicts the annual distribution of bibliographic records of journal papers on embodied carbon from 1990 to 2024. Up until 2014, fewer than ten papers were published each year globally. Notably, a surge in publication volume commenced around 2014, with the number of papers published annually growing at an exponential rate.

In total, the 372 publications were published in 109 different journals. This high number of journals indicates a wide variety of disciplines involved. Those journals can be classified into the following categories: construction building

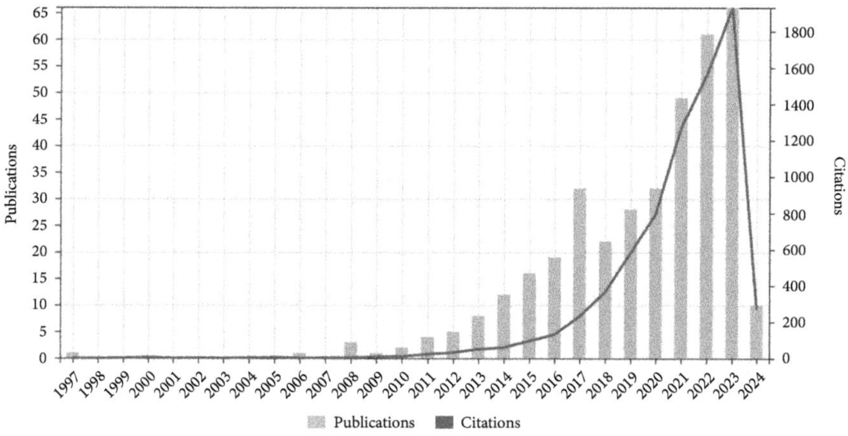

Figure 1.2 Distribution of published peer-reviewed journal paper. Courtesy of Ming Hu.

Source: data from Web of Science, http://www.webofscience.com

technology, green sustainable science technology, environmental engineering, civil engineering, environmental science, material science, energy engineering, and architecture.

The author then conducted a co-citation analysis to gain an overview of the research focus areas. The more frequently two publications are cited together in other works, the stronger their intellectual relationship and shared contribution. In the map depicted in Figure 1.3, each circle represents a paper, and those of the same colour (black, dark grey, light grey) indicate a common topic among them. The proximity of two circles signifies the strength of the intellectual connection, suggesting similar contributions to a particular research focus or an emerging research trend. The larger the circle, the more citations the paper has received. The three main clusters identified are (black) embodied energy components and mitigation strategies, (dark grey) the trade-off between energy efficiency and environmental impact, and (light grey) regional analysis and significance of embodied energy.

Cluster One (black) represents the evolution of knowledge regarding the constituents of embodied carbon calculations and associated mitigation strategies. Cluster Two (dark grey) concentrates on the analysis of actual projects and a range of environmental impact categories. The close relationship between the first and second clusters highlights the intersecting research interests and activities between the fields of building construction technology and environmental science. The findings from Cluster Two also validate the results from Cluster One concerning the proportion of embodied energy use in the

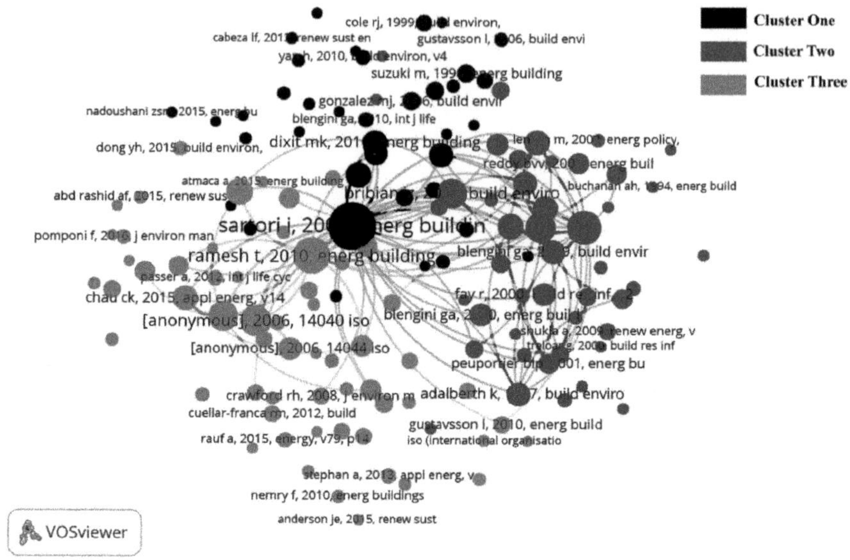

Figure 1.3 Research topic analysis. Courtesy of Ming Hu using VOSviewer software.
Source: data from Web of Science, http://www.webofscience.com

total life-cycle energy consumption. Depending on the region and climatic conditions, embodied energy constitutes approximately 10–27% of the total energy use and carbon emissions for conventional commercial buildings [36, 37, 38]. 15–40% for residential buildings [39], 26–60% for low-energy use buildings [38, 40], and 74–100% for nearly net-zero energy buildings [41]. In a simple way to describe it the more energy-efficient the building operation is, the higher the percentage embodied energy counts from the life cycle of a building. Wooden structures are shown to consume less embodied energy than steel or concrete structures, with a differential of 27–60% [42]. Furthermore, a building's structural system is responsible for a significant portion of its overall embodied energy, roughly 50% [42]. The third cluster (light grey) encapsulates recent research concentrating on regional analysis on the importance of embodied carbon and energy.

1.5.2 Country and regional interest

As of March 2024, knowledge on embodied carbon, as evidenced by peer-reviewed publications, spans seventy-two countries, with Europe contributing twenty-seven countries, Asia thirty, South America three, North America three, Oceania two, and Africa seven. Figure 1.4 displays the global distribution of contributing countries and regions, revealing that sixty-three countries (87.5%) have produced ten or fewer publications, whereas nine countries and regions (12.5%) have produced more than ten papers. The United Kingdom leads with sixty-three publications, followed by China with fifty-nine, the United States with fifty-three, and Australia with forty-nine. Other countries with more than ten publications include Italy with twenty-two, Malaysia with twenty, India with fourteen, Scotland with fourteen, and Belgium with twelve.

Embodied carbon research has become a global pursuit, with both developing and developed countries contributing substantially to the discourse in recent decades. Global research activities and collaborations have transcended the economic development status of regions as all countries grapple with environmental risks and challenges posed by the building industry. From the knowledge generation distribution, it is clear that geopolitical or economic status does not appear to be a consistent driver of scientific and academic output, as the top five countries represent a mix of Global South and Global North nations. It is very promising to see that the research efforts are geographically diverse, with the three most prolific countries situated on separate continents: Europe, North America, and Asia. As outlined in Section 1.5.1, an accurate and holistic assessment of energy flow and associated embodied carbon requires a reliable system boundary definition. In the current era, because of the intricate supply chain, assessment of the actual embodied carbon of a product may need wider and cross-continent system boundaries. Hence, contributions from various countries and regions are imperative for propelling research forward in this relatively nascent field.

What is more interesting is that there are different focuses and interests in knowledge-seeking between the Global South (developing countries) and the Global North (developed countries). In this sense, socioeconomic factors have affected the trend of embodied carbon research. The Global South concentrates more on methodologies, procedures, and the accuracy of embodied carbon and energy quantification. In contrast, the Global North is more invested in striking a balance between energy efficiency and environmental impact, acknowledging that their total energy use is expanding at a slower pace than that in the Global South.

The terms Global North and Global South are used to categorize countries based on their economic development and political power. The Global North typically includes economically developed regions such as Europe, North America, Australia, Israel, South Africa, and parts of Asia (e.g. Japan), while

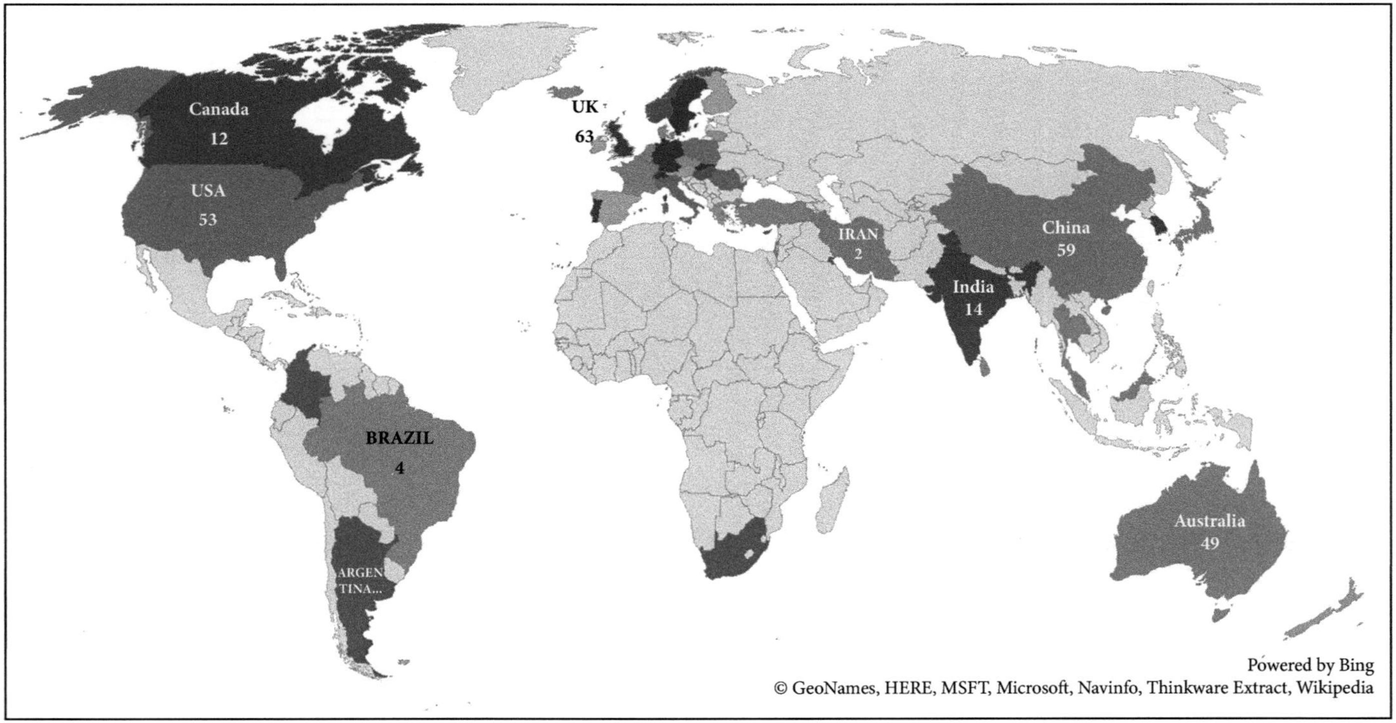

Figure 1.4 Number of publications distributed by country and region. Courtesy of Ming Hu.
Source: data from Web of Science, http://www.webofscience.com

the Global South comprises economically developing countries from regions like Africa, India, China, Brazil, Mexico, and Latin America [43, 44]. The terms Global North and Global South gained popularity when the Brandt Commission utilized them in the 1980s to illustrate the global division between wealthier countries in the northern hemisphere and less affluent nations in the southern hemisphere. This will be explained in detail in Chapter 5. Although initially associated with hemispheric distinctions, the terms have evolved beyond clear-cut geographical definitions [45]. These terms are not merely geographical but also reflect disparities in wealth and influence; the terms highlight the stark economic disparities and power differentials between regions, emphasizing the divide between richer and poorer nations [45]. The distinction between the Global North and Global South is significant in various fields, including migration studies [46], global health collaborations [46], and climate change negotiations. Climate change discussions often involve negotiations between the industrialized Global North and the rest of the world, particularly the Global South [47].

Moreover, the Global North and Global South classifications have implications for research methodologies and knowledge production. Studies in the Global South sometimes rely on concepts and frameworks originating from the Global North, indicating a need to develop theories grounded in southern contexts [48]. The dynamics between these two global regions also influence academic collaborations, with efforts to bridge the gap between different knowledge systems and scholarly dialogues [49].

In the case of embodied carbon research, these differences may stem from the Global South's dual struggle with environmental degradation and energy poverty. In these developing nations, overall energy consumption is set to rise despite advancements in building energy efficiency. By 2050, it is projected that China's total energy consumption in the building sector will be fifteen times greater than in 1970. Similarly, Brazil and India are expected to experience an eleven- to twelve-fold increase in energy consumption by 2050 compared to 1970 [50], coinciding with population growth and rising living standards. In 2016, 18% of India's population and 27% of the population in other developing Asian countries lacked access to electricity [51], an issue that must be addressed for these countries to progress. With projected high population growth and increased energy consumption accompanying improved living standards, the environmental impact is set to become a major concern for the Global South in the upcoming decades. In response, developing countries are earnestly seeking or adopting low-energy, low-carbon, and low-impact developmental paths that concurrently promote energy conservation and environmental impact reduction. Therefore, robust and reliable knowledge is needed to inventory, measure, and predict energy use and embodied carbon increase. Then, mitigation and reduction can be planned. That explains why Global South focuses on the method, procedure, prediction, and precision.

Conversely, the Global North's developed countries are contending with ageing building stock and infrastructure. In the European Union, for instance, 75% of buildings were constructed before 1990 and could be more energy-efficient, necessitating significant upgrades [52]. Therefore, retrofitting building stocks has become a critical priority for both Europe and other developed nations [53]. Renovations and energy retrofits can conserve the embodied energy within existing structures, thereby prioritizing the need for precise methodologies to measure embodied energy and carbon and its contribution to climate change mitigation. A crucial first step towards an accurate assessment of embodied energy is to establish an optimization framework, which has been developed in numerous developed countries over the past few decades.

1.5.3 Temporal shift in knowledge creation

There has been a temporal shift in embodied carbon research fields and an expansion in the scope of research topics, as illustrated in Figure 1.5.

In the early 1990s (1990–5), two primary disciplines—building construction and environmental science—significantly contributed to the development and advancement of embodied carbon research. Over a decade, the interest and impact in embodied carbon broadened to include disciplines such as energy fuels, ecology, materials science, urban studies, and public policy [54].

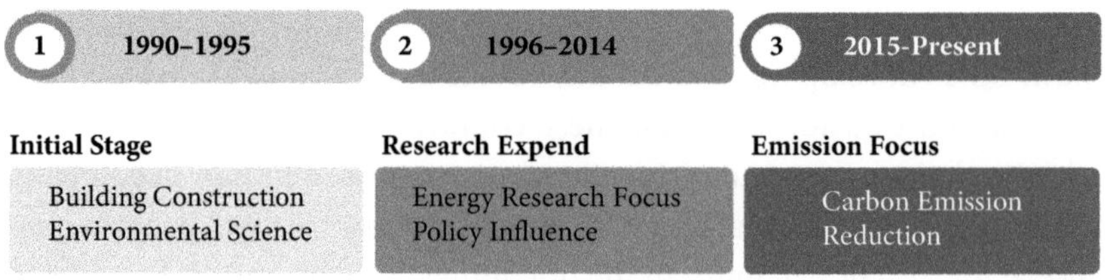

Figure 1.5 Temporal shift in knowledge creation around embodied carbon research. Courtesy of Ming Hu.

Between 1996 and 2014, research concentrated on environmental problems stemming from energy-inefficient building designs, with evidence suggesting that operating energy was a primary contributor to environmental degradation. The U.K.'s policy decision was based on this. Following this realization, strategies for mitigation were proposed, including adding insulation to the building envelope, which yielded immediate improvements in energy conservation and reductions in environmental impact but increased embodied carbon [54].

Beginning in 2015, reducing environmental impacts, particularly carbon emissions, became a priority, uniting researchers across various fields. This shift led to re-evaluating the balance between operating and embodied energy and carbon, approaching mitigation strategies from a holistic life-cycle perspective. The full spectrum of embodied carbon, including that in construction processes, transportation, and demolition, came under focus [54].

Since 2017, factors beyond the realms of building construction and environmental science have been gradually incorporated, accounting for uncertainties buildings may face over their extended lifespan. As operating energy efficiency continues to improve globally, the significance of embodied energy is increasing. Research efforts are intensifying in the search for alternative materials to traditional energy-intensive options, such as concrete, which contribute significantly to environmental impacts [2].

1.6 Embodied carbon and climate justice

As discussed, the interest and investment in understanding the inclusion of embodied carbon in the built environment is increasing. The concept of embodied carbon becomes crucial in this discourse, particularly because it encapsulates the carbon emissions associated with the entire life cycle of the built environment—from the extraction and processing of raw materials to the construction, maintenance, and eventual demolition of buildings. The built environment, including roads, railways, housing, schools, factories, and office towers, is a tangible manifestation of these emissions. As such, the historical accumulation of carbon emissions is directly linked to the physical structures that define human settlements and economies, reflecting a legacy of environmental impact that is deeply interwoven with the fabric of modern society.

This historical accumulation of emissions and the resultant climate change have raised critical questions about climate justice. There is an emerging recognition of the need to address not only the inequities in emissions contributions but also the disparities in vulnerability and capacity to respond to climate impacts. The Global South, despite contributing less to historical emissions, faces more severe impacts of climate change and often has limited resources to adapt to these changes or to transition to low-carbon development pathways.

In the contemporary context, the construction sector and its supply chain play a significant role in exacerbating the inequitable distribution of embodied carbon emissions. The global supply chain for building materials such as steel and cement is a significant source of carbon emissions, accounting for 6.5 and 7.0% of global carbon emissions, respectively [55]. As the demand for infrastructure and urban development continues to rise, particularly in emerging economies in the Global South, the challenge of managing embodied carbon becomes increasingly complex. This situation is further complicated by the fact that the environmental impacts of these emissions disproportionately affect the most vulnerable populations and communities, who are often the least responsible for historical emissions.

To address these challenges, there is a growing call for sustainable construction practices that minimize embodied carbon through innovative design, material selection, and construction techniques. Additionally, policies and frameworks that promote carbon accounting and life-cycle assessment in the built environment are critical for making visible the hidden costs of embodied carbon. Ultimately, achieving climate justice in the context of the built environment requires a concerted effort that includes technological innovation, policy reform, and international cooperation to ensure a fair and equitable transition to a low-carbon future.

1.7 Moving forward

The journey of embodied carbon is a fascinating tale of scientific curiosity, policy evolution, and practical application. From early concepts of embodied energy in the late nineteenth century to today's sophisticated methodologies, our understanding and management of embodied carbon have become pivotal in combating climate change. Visionaries like Frederick Soddy, Nicholas Georgescu-Roegen, and Herman Daly have been instrumental in this evolution, advocating for the integration of ecological principles into economic and environmental strategies. Initially met with resistance and fluctuating urgency, the importance of embodied carbon is now firmly recognized, with decades of research and development building a strong foundation for current and future efforts to minimize the carbon footprint of our built environment.

This book charts a new course in understanding embodied carbon issues and the related inequities between the Global North and South. It challenges entrenched norms, rectifies biases, and inspires a more balanced and responsible approach to mitigating carbon emissions in the built environment. The core arguments emphasize that persistent global inequality, ongoing development needs in the Global South, and the imbalanced suffering, responsibility, and action on carbon emissions are the greatest threats to future North–South policy cooperation. The root problem lies in the ignorance of the true value of embodied carbon and the bias towards continuous growth and technological advancement, often driven by unilateral carbon-neutral requirements set by Global North institutions. Addressing these issues is crucial to preventing the world from moving further away from carbon neutrality.

The book comprises three parts. Part I, 'Looking Back and Perception', explores the historical evolution of embodied carbon, highlighting its significance in the context of climate change and its role in climate action. Part II, 'Unravelling the Complexity and the Hidden Inequities', examines the environmental justice perspective, the significant burden shifts to the Global South, and the biases of high-tech solutions. Part III, 'Environmental Equality for the Future', investigates innovative, equitable, and sustainable strategies to reduce embodied carbon, emphasizing low-tech alternatives and insights from regional and traditional practices. Each part is related but independent, allowing readers to read sequentially or jump to the parts or chapters that interest them the most.

Part I includes Chapters 1, 2, and 3. Chapter 1 underscores the global effort, with valuable contributions from both the Global North and Global South, each offering unique insights and solutions. Chapter 2 delves into the definitions, components, and accounting methods of embodied carbon, unravelling the complexities of the environmental product declaration market. Chapter 3 connects embodied carbon to climate change, highlighting its significance in climate action and real-world impacts, laying the groundwork for understanding the foundational aspects of embodied carbon and its crucial role in global carbon emissions.

Part II includes Chapters 4 through 6. Chapter 4 explores the environmental justice perspective on embodied carbon, highlighting the significant burden shifts to the Global South, the paradoxes of energy use and emissions, and the inequities hidden within the production and consumption of green technologies. Chapter 5 examines the Global North bias in sustainability, showing how it marginalizes the Global South by skewing policies and metrics in favour of affluent nations. Chapter 6 critiques the high-tech bias in sustainability, emphasizing the limitations and negative impacts of advanced technologies while advocating for effective low-tech solutions like material reuse and efficient design for reducing embodied carbon.

Part III includes Chapters 7 through 9. Chapter 7 reviews current regulations and policies, while Chapter 8 proposes a low-tech approach based on traditional and indigenous practices. Chapter 9 presents insights from multidisciplinary experts, advocating for a comprehensive approach to reset the current imbalance and create equitable solutions to reduce embodied carbon emissions and their impact on planetary health.

References

1. Mohebbi G, Hasan A, Blay-Armah A, Bahadori-Jahromi A, Mylona A, Barthorpe M. Comparative analysis of the whole life carbon of three construction methods of a UK-based supermarket. Building Services Engineering Research & Technology 2023; 44(3): 355–75.
2. Hu M, Esram NW. The status of embodied carbon in building practice and research in the United States: a systematic investigation. Sustainability 2021 Nov 23; 13(23): 12961.
3. United Nations Environment Programme. 2021 Global status report for buildings and construction—towards a zero-emissions, efficient and resilient buildings and construction sector. Nairobi; 2021.
4. Hawkins D, Mumovic D. Evaluation of life cycle carbon impacts for higher education building redevelopment: a multiple case study approach. Energy and Buildings. 2017; 150: 507–15.
5. Gütschow J, Jeffery ML, Gieseke R, Gebel R, Stevens D, Krapp M, et al. The PRIMAP-hist national historical emissions time series. Earth System Science Data. 2016; 8(2): 571–603.
6. House of Commons. Building to net zero: costing carbon in construction [Internet]. 2022 Nov [Accessed 21 May 2024]. Available from https://publications.parliament.uk/pa/cm5803/cmselect/cmenvaud/103/report.html
7. Pomponi F, D'Amico B, Moncaster A. A method to facilitate uncertainty analysis in LCAs of buildings. Energies. 2017 Apr 13; 10(4): 524.
8. Zhang B, Bai S, Ning Y. Embodied energy in export flows along global value chain: a case study of China's export trade. Front Energy Res. 2021 May 4; 9: 649163.
9. Hu M. A building life-cycle embodied performance index—the relationship between embodied energy, embodied carbon and environmental impact. Energies. 2020; 13(8): 1905.

10. U.S. Energy Information Administration. Carbon dioxide emissions coefficients [Internet]. Environment. 2023 [Accessed 12 Mar 2024]. Available from https://www.eia.gov/environment/emissions/co2_vol_mass.php
11. U.S. Environmental Protection Agency. Power Profiler. [Internet]. eGRID. [Accessed 12 Mar 2024]. Available from https://www.epa.gov/egrid/power-profiler#/
12. Costanza R. Embodied energy and economic valuation. Science. 1980 Dec 12; 210(4475): 1219–24.
13. Spash CL. Routledge handbook of ecological economics. New York:Routledge; 2017.
14. Hu M. Net zero energy building: predicted and unintended consequences. New York:Routledge; 2019.
15. Liu Z, Koerwer J, Nemoto J, Imura H. Physical energy cost serves as the 'invisible hand' governing economic valuation: direct evidence from biogeochemical data and the U.S. metal market. Ecological Economics. 2008 Aug; 67(1): 104–8.
16. Kapitonov IA, Voloshin VI, Korolev VG. Energy security of territories as a factor of sustainable development under the conditions of economic changes. IJEEP. 2019 Oct 1; 9(6): 210–21.
17. Demaria F, Schneider F, Sekulova F, Martinez-Alier J. What is degrowth? From an activist slogan to a social movement. Environmental Values. 2013 Apr 1; 22(2): 191–215.
18. Gowdy J, Mesner S. The evolution of Georgescu-Roegen's bioeconomics. Review of Social Economy. 1998 Jun; 56(2): 136–56.
19. Veraart R, Blok V. Towards a philosophy of a bio-based economy: a Levinassian perspective on the relations between economic and ecological systems. Environmental Values. 2021 Apr; 30(2): 169–92.
20. The History of Economic Thought. Nicholas Georgescu-Roegen, 1906-1994 [Internet]. [Accessed 12 Mar 2024]. Available from https://www.hetwebsite.net/het/profiles/georgescu.htm
21. Rammelt C. The spiralling economy: connecting Marxian theory with ecological economics. Environmental Values. 2020 Aug; 29(4): 417–42.
22. Kish K, Farley J. A research agenda for the future of ecological economics by emerging scholars. Sustainability. 2021 Feb 2; 13(3): 1557.
23. Daly HE. Ecological economics and sustainable development: selected essays of Herman Daly. Cheltenham, UK; Northampton, MA: Edward Elgar; 2007.
24. Zhu D, Zhang S, Sutton DB. Linking Daly's proposition to policymaking for sustainable development: indicators and pathways. Journal of Cleaner Production. 2015 Sep; 102: 333–41.
25. O'Neill DW. Herman E. Daly (1938–2022). Nat Sustain. 2022 Dec 29; 6(2): 118–19.
26. Marzetti GV. The fund-flow approach: a critical survey. Journal of Economic Surveys. 2013 Apr; 27(2): 209–33.
27. Gowdy JM, Iorgulescu Polimeni R. The death of *homo economicus*: is there life after welfare economics? International Journal of Social Economics. 2005 Nov 1; 32(11): 924–38.
28. Georgescu-Roegen N, Leontief WW. The structure of the American economy, 1919-1939: an empirical application of equilibrium analysis. Econometrica. 1951 Jul; 19(3): 351.

29. Herendeen RA. Energy cost of goods and services [Internet]. 1973 Oct [Accessed 11 Mar 2024] p. ORNL-NSF-EP-58, 4432615. Report No. ORNL-NSF-EP-58, 4432615. Available from http://www.osti.gov/servlets/purl/4432615/
30. Herendeen R, Tanaka J. Energy cost of living. Energy. 1976; 1(2): 165–78.
31. Herendeen RA. Notes and comments on the energy cost of building construction. Energy and Buildings. 1977 May; 1(1): 95–6.
32. Bullard CW, Penner PS, Pilati DA. Net energy analysis: handbook for combining process and input-output analysis. Resources and Energy. 1978 Nov; 1(3): 267–313.
33. Buck AL. History of the Energy Research and Development Administration [Internet]. 1982 Mar [Accessed 11 Mar 2024] p. DOE/ES-0001, 5272047. Report No. DOE/ES-0001, 5272047. Available from http://www.osti.gov/servlets/purl/5272047/
34. Odum HT, Brown M, Brandt-Williams S. Handbook of energy evaluation. Center for Environmental Policy. 2000; 45.
35. Azari R, Moncaster A, editors. The Routledge handbook of embodied carbon of the built environment. Abingdon, Oxon: Routledge, Taylor & Francis Group; 2024.
36. Chen C. CiteSpace: a practical guide for mapping scientific literature. Hauppauge, NY:Nova Science Publishers; 2016.
37. Pomponi F, Moncaster A. Embodied carbon mitigation and reduction in the built environment–what does the evidence say? Journal of Environmental Management. 2016; 181: 687–700.
38. Van Eck NJ, Waltman L. Visualizing bibliometric networks. In: Measuring scholarly impact: methods and practice. New York: Springer; 2014, pp. 285–320.
39. Giesekam J, Pomponi F. Briefing: embodied carbon dioxide assessment in buildings: guidance and gaps. London: Thomas Telford; 2017, pp. 334–41.
40. Stephan A, Crawford RH, De Myttenaere K. Multi-scale life cycle energy analysis of a low-density suburban neighbourhood in Melbourne, Australia. Building and Environment. 2013; 68: 35–49.
41. Birgisdottir H, Moncaster A, Wiberg AH, Chae C, Yokoyama K, Balouktsi M, et al. IEA EBC annex 57 'evaluation of embodied energy and CO2eq for building construction'. Energy and Buildings. 2017; 154: 72–80.
42. Pomponi F, Piroozfar PA, Southall R, Ashton P, Farr ER. Life cycle energy and carbon assessment of double skin façades for office refurbishments. Energy and Buildings. 2015; 109: 143–56.
43. Kamuya D, Bitta MA, Addissie A, Naanyu V, Palk A, Mwaka E, et al. The Africa Ethics Working Group (AEWG): a model of collaboration for psychiatric genomic research in Africa. Wellcome Open Res. 2021 Jul 27; 6: 190.
44. Odeh LE. A comparative analysis of global north and global south economies. Journal of Sustainable Development in Africa. 2010; 12(3): 338–48.
45. Seo H, Thorson S, Blomberg M, Appling S, Bras A, Davis-Roberts A, et al. Country characteristics, internet connectivity and combating misinformation: a network analysis of Global North–South. In 2021 [Accessed 15 Mar 2024]. Available from: http://hdl.handle.net/10125/70975
46. Czaika M, De Haas H. The globalization of migration: has the world become more migratory? International Migration Review. 2014 Jun; 48(2): 283–323.

47. Chen Y. How has ecological imperialism persisted? A Marxian critique of the Western climate consensus. American J Econ Sociol. 2022 May; 81(3): 473–501.
48. Katsui H, Swartz L. Research methods and practices of doing disability studies in the Global South. Scandinavian Journal of Disability Research. 2021 Jul 9; 23(1): 204–6.
49. Van Breda ADP, Pinkerton J. Raising African voices in the global dialogue on care-leaving and emerging adulthood. Emerging Adulthood. 2020 Feb; 8(1): 6–15.
50. International Atomic Energy Agency. Energy, electricity and nuclear power estimates for the period up to 2050 [Internet]. Vienna; 2020 [Accessed 15 Mar 2024]. Available from https://www.pub.iaea.org/MTCD/Publications/PDF/RDS-1-40_web.pdf
51. Organisation for Economic. Co-operation and development. Energy Access Outlook 2017: From poverty to prosperity [Internet]. 2017 Oct. Available from https://www.oecd.org/publications/energy-access-outlook-2017-9789264285569-en.htm
52. Buildings Performance Institute Europe. State of the building stock briefing [Internet]. Available from https://bpie.eu/wp-content/uploads/2017/10/State-of-the-building-stock-briefing_26Ott_v1.pdf
53. Pombo O, Rivela B, Neila J. The challenge of sustainable building renovation: assessment of current criteria and future outlook. Journal of Cleaner production. 2016; 123: 88–100.
54. Hu M, Milner D. Visualizing the research of embodied energy and environmental impact research in the building and construction field: A bibliometric analysis. Developments in the Built Environment. 2020 Aug; 3: 100010–100010.
55. Fennell P, Driver J, Bataille C, Davis SJ. Cement and steel—nine steps to net zero. Nature. 2022; 603(7902): 574–7.

Chapter 2
Define embodied carbon

2.1 Current definitions of embodied carbon

In Chapter 1, we provided a large picture of the evolution of embodied carbon. After a century of development, the concept of embodied carbon has been accepted into the mainstream conversation regarding climate change mitigation and life-cycle consideration. However, in the built environment, the concept and assessment of embodied carbon remain ambiguous, sometimes even ill-defined, leading to inaccurate comparisons and assessments. Let us first demystify embodied carbon in the built environment, delineating its roles and forms from a whole life-cycle perspective.

Picture this: embodied carbon as the invisible backpack of carbon dioxide emissions that materials carry from extraction to demolition. As illustrated in Figure 2.1, these emissions can be broken down into different life-cycle stages, marked A through C. Let's unpack this backpack to understand what each part contains.

In the full life cycle of a building product or building, Stage A is where the story unfolds—from the raw material extraction (A1) right up to the building's construction phase (A5). It's here we set the scene for embodied carbon, which can be measured in distinct acts:

Embodied Carbon, Act One: 'Cradle to Gate (EC1)', which takes us from material birth (A1) to the factory gate (A3). It's like following a product's backstory without seeing it take on its role in the world.

Embodied Carbon, Act Two: 'Cradle to Site (EC2)', expanding the narrative through stage A5, tracking the journey until materials reach their construction stage. It's a deeper dive, yet the story isn't quite complete.

Embodied Carbon, the Full Story: 'Cradle to Grave (EC3)', encompassing the entire life cycle of a material, from its initial creation (stage A) to its ultimate disposal (stage C). This life cycle represents the total carbon footprint of a building, while operational carbon (stages B6 and B7) reflects the emissions generated during the building's use, much like its ongoing energy consumption. A building's carbon emissions

Embodied Carbon. Ming Hu, Oxford University Press. © Ming Hu (2025). DOI: 10.1093/9780198944485.003.0002

Building Life Cycle Stage					
Product	A1	Raw material extraction	Cradle to Gate	Cradle to Site	Cradle to Grave
	A2	Transport			
	A3	Manufacturing			
Construction	A4	Transport			
	A5	Construction/Installation			
Use	B1	Use			
	B2	Maintenance			
	B3	Repair			
	B4	Replacement			
	B5	Refurbishment			
	B6	Operational energy use			
	B7	Operational water use			
End of Life	C1	Deconstruction			
	C2	Transport			
	C3	Waste processing			
	C4	Disposal			

Figure 2.1 Embodied carbon definition and life-cycle stages.

can be broadly categorized into two types: life-cycle embodied carbon and operational carbon. The former covers all carbon emissions associated with material extraction (module A1), transport to manufacture (A2), manufacturing (A3), transport to site (A4), construction (A5), use phase (B1, e.g. concrete carbonation), maintenance (B2), repair (B3), replacement (B4), refurbishment (B5), deconstruction (C1), transport to end-of-life facilities (C2), waste processing (C3), and eventually disposal (C4) [1, 2]. In contrast, operational carbon (B6 and B7) refers explicitly to emissions from the energy required to heat, cool, use water, and power the building.

Figure 2.2 presents a building stock analysis of the ratio between operational carbon and embodied carbon in Denver, Colorado, over 40 years. As building codes progress towards net-zero energy and decarbonization of heating through electrification, the share of operational carbon in a building's total emissions decreases. For instance, in Denver, embodied carbon is projected to account for 27% of all building-related emissions by 2030, compared to the annual operational emissions for the same period. If Denver achieves its target of all buildings being net-zero energy by 2040, 100% of emissions will come from embodied carbon.

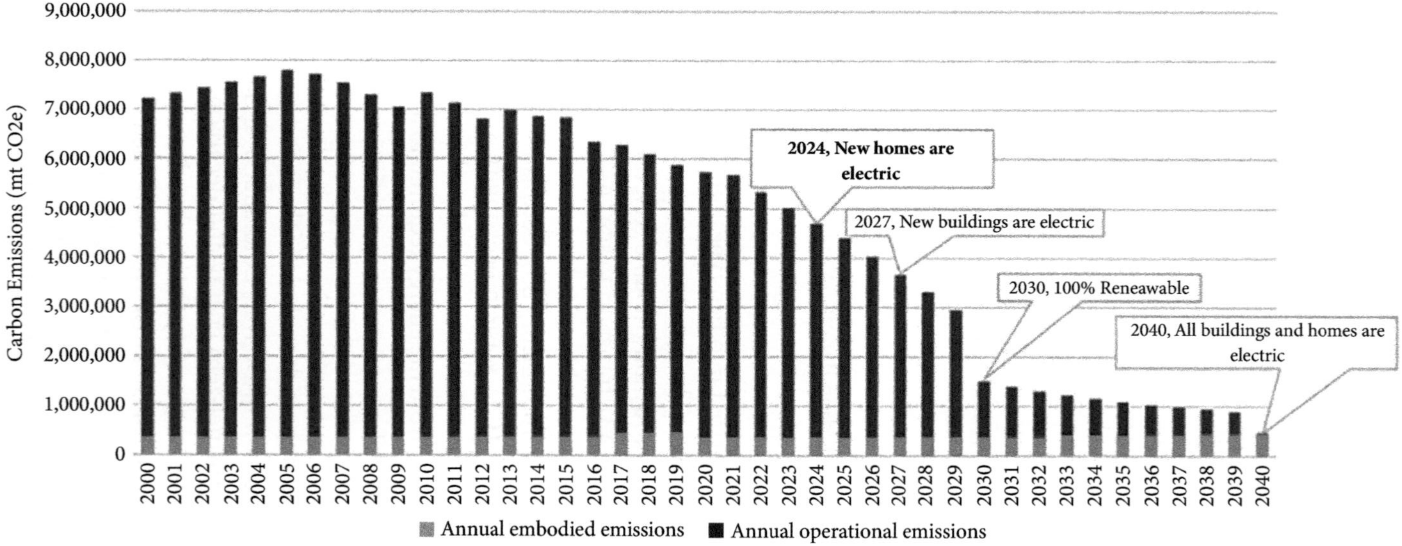

Figure 2.2 Embodied carbon and operational emissions for all buildings by year.

Reproduced with permission from City and County of Denver Office of Climate Action, Sustainability, and Resiliency and Community Planning and Development, United States. Anderson et al., 'Denver's Building Sector Embodied Carbon Emissions' (June 2011).

When choosing materials, this distinction of different embodied carbon acts is critical. Consider two insulated windows: one claims a lower carbon count with EC2, and the other is seemingly higher with EC3. A budget-conscious student or professional might lean towards the EC2 option if costs align. But is it an apples-to-apples comparison? Without considering the full life cycle (EC3), we might miss crucial acts in our material's drama. So, in our quest for sustainability, we must ask: have we chosen the most sustainable product, or have we been swayed by an incomplete story?

For those shaping the future of construction, this isn't just about numbers—it's about making informed, fair choices in the materials we use. It's a call to ensure that our environmental impact calculations are not just precise but also all-encompassing. Now that we understand embodied carbon is associated with every life stage of a building, let us look into the global trend of the building and construction sector.

2.2 How is embodied carbon measured?

The measurement of embodied carbon is closely linked to the development of life-cycle assessment (LCA) methodologies for materials and buildings. Embodied carbon is typically assessed as part of the overall life-cycle carbon footprint of a building project. LCA is a systematic method used to evaluate the environmental impacts of a product or service throughout its entire life cycle, [3]. An LCA is composed of four main stages: defining the goal and scope (establishing the objectives and boundaries of the assessment), conducting an inventory analysis, performing an impact assessment, and interpreting the results. The International Organization for Standardization (ISO) formalized the LCA method in the 1990s with standards ISO 14040 and ISO 14044, which are still the most widely used versions today and are regularly updated [4]. In addition, ISO 21930 and ISO 21931 provide further guidance specifically for applying LCA in the construction industry, at both the product and building levels [2].

When LCA is applied in the building and construction sector, assessments can be conducted at four different levels, depending on the chosen scope, system boundaries, and functional units: the building material level, the building product level (including components, systems, and assemblies), the whole building level, or the industry level. Despite these different scopes, all levels typically rely on the same database of inputs. Among these, LCAs are most commonly performed at the material and whole building levels. For an LCA to be effective, the project team must clearly define the goal and scope at the outset and determine which level of LCA is appropriate for the project.

Throughout each stage of a building's life cycle—from cradle to grave—the building interacts with the environment by consuming raw materials and producing carbon emissions and pollutants. LCA provides a quantitative framework to describe these interactions and their associated environmental and human health impacts, including metrics such as global warming potential (GWP), ozone depletion potential (ODP), eutrophication potential (EP), acidification potential (AP), and smog formation potential (SFP).

Embodied carbon quantification is specifically derived from the GWP metric, which measures various greenhouse gas emissions (such as methane, nitrous oxide, and chlorofluorocarbons) in terms of their CO_2 equivalence. Life-cycle embodied carbon can be further broken down into three primary categories listed in Box 2.1: upfront embodied carbon [1], recurring embodied carbon, and end-of-life embodied carbon [2].

Box 2.1: Life-cycle embodied carbon categories [2]

Upfront embodied carbon (UEC) represents the emissions generated both on-site and off-site before a building becomes operational. This includes activities such as material production, construction, prefabrication, transportation, and site preparation, corresponding to life-cycle stages A1 to A5. UEC often makes up a significant portion of the total life-cycle carbon and can be reduced through strategies like minimizing building size, using less energy-intensive materials, and optimizing design choices (e.g. preferring lightweight structural systems over heavier alternatives).

Recurring embodied carbon (REC) accounts for the emissions produced during the building's use phase, related to repair, replacement, refurbishment, and maintenance activities. REC covers life-cycle stages B1 to B5.

End-of-life embodied carbon (EEC) includes emissions generated during demolition, deconstruction, transport of deconstructed materials, waste processing, and disposal, spanning life-cycle stages C1 to C4.

Both upfront embodied carbon (UEC) and end-of-life embodied carbon (EEC) occur only once over a building's lifetime, while recurring embodied carbon (REC) can fluctuate significantly based on factors such as location, usage conditions, and maintenance frequency. Previous studies have identified UEC (stages A1–A5) as a major contributor to total embodied carbon emissions. For instance, Liu et al. (2020) and Xie et al. (2015) found that phases A1–A3 alone can account for approximately 40 to 45% of a building's total life-cycle embodied carbon emissions [5, 6]. Additionally, Blay-Armah et al. (2022) reported that UEC (A1–A5) represents over 55% of total life-cycle embodied carbon, with REC contributing around 20% and EEC only about 2% of emissions [7]. By understanding these categories and their respective stages, project teams can more effectively measure and mitigate a building's embodied carbon, supporting the adoption of more sustainable construction practices.

2.3 Variability and influencing factors of embodied carbon emissions

Research into the carbon footprint of building materials has highlighted the significant variation in embodied carbon emissions, reflecting the diversity of construction practices and materials used globally. Early research focused on the embodied energy of foundational materials like steel and concrete, extending to more specific components such as precast concrete wall panels. Studies at the building Level are predominantly case-specific. For instance, a notable study on a three-story office edifice in India recorded total life-cycle carbon emissions at 450 $kgCO_2e/m^2$ [8], a figure that starkly contrasts with findings from other countries—ranging from 930 $kgCO_2e/m^2$ in Thailand to as high as 3340 $kgCO_2e/m^2$ in Italy [6, 7]. The breadth of these findings is mirrored in an extensive 2018 review by Chastas and colleagues., which synthesized data from ninety-five case studies worldwide, including a notable proportion from the United States [9]. This review revealed that residential buildings' normalized embodied carbon emissions fluctuate widely, from 179.3 to 1050 $kgCO_2e/m^2$. Moreover, embodied carbon can represent 9 to 80% of a building's total life-cycle carbon emissions [9]. These disparities stem largely from variations in the studies' system boundaries, the assumed lifespan of the buildings, and the sources of the data they utilized [10, 11].

The significant discrepancies in building carbon emissions are not solely attributable to geographical diversity; other contributing factors include local climatic conditions and the type of building construction. The following subsections will first provide an overview of construction materials and then delve into two pivotal aspects: the production of materials and the maintenance requirements that epitomize life-cycle stages A and B.

2.3.1 Where does the material come from? Life-cycle stage A1

The term 'construction material' encompasses any substance utilized in the creation of buildings and infrastructure. Historically, structures have risen from the earth, crafted from various natural materials, such as pebbles, clay, wood, sand, twigs, and leaves. Today, the industry relies on a more diverse and nonorganic palette, ranging from bricks and timber to innovative materials like corrosion-resistant additives, polymers, recycled content, glass fibres, concrete, cement, and advanced composites. The expansion of the construction material market is primarily propelled by its escalated consumption in commercial and residential development. The engine driving this growth encompasses several key factors: burgeoning infrastructure initiatives, demographic expansion, rising individual incomes, accessible loan facilities, and supportive governmental policies.

After a slight contraction in 2022, global construction saw a 0.9% uptick in 2023, signalling a rebound from pandemic-induced setbacks [12]. Projections suggest a steady 3% growth in construction activity from 2025 to 2030, a rate that mirrors global GDP growth trends. Countries like the Philippines, India, and Bangladesh are leading in growth rates, with mature economies like the United States close behind. Japan, however, is expected to see a decrease in construction output [12]. Housing remains a strong sector, with India expected to average over 12 million new starts annually from 2024 to 2030. This is bolstered by governmental housing schemes across various nations.

In the Global North developed countries, the new construction growth is predicted to be sluggish in Western Europe. For example, Belgium and France seeing modest increases of 0.7 and 0.8% per annum, respectively [13]. But, like in Global South countries, the housing sector in the Global North is a bright spot. For example, the New South Wales government introduced the Property Tax (First Home Buyer Choice) Bill 2022, which offers first-time homebuyers a novel tax-paying option to replace the traditional stamp duty on residential property purchases. This legislative move is projected to support an estimated 97% of first-home buyers, benefitting approximately 57,000 individuals annually [14]. In developed countries, the renovation and modernization of ageing infrastructure and housing will likely surge the demand for construction materials. The U.S. economy, for instance, with its $20 trillion value, is reliant on the renovation of its vast infrastructure network, with Congress allocating $1.2 trillion for such improvements [15]. Yet, the sourcing and final destination of

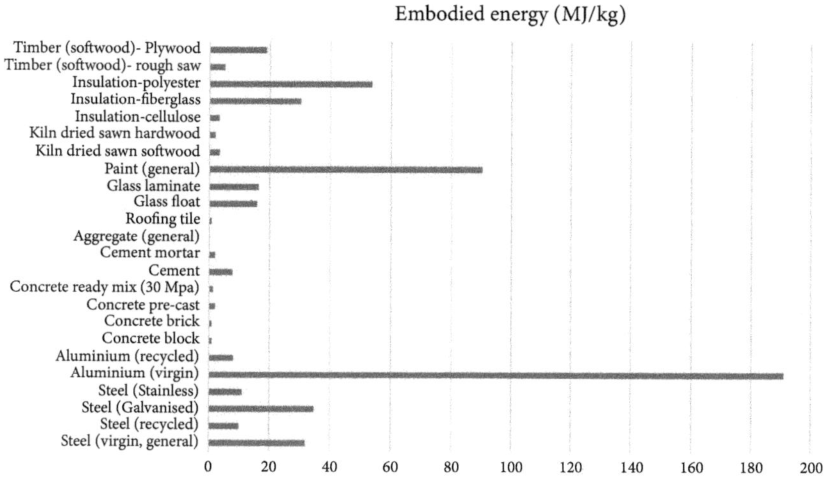

Figure 2.3 Embodied energy intensities for common building materials.

Source: data from Alcorn, A. (2001). Embodied energy and CO2 coeffects for New Zealand building materials. Wellington, New Zealand: Centre for Building Performance Research, Victoria University of Wellington. https://www.wgtn.ac.nz/architecture/centres/cbpr/resources/pdfs/ee-co2_report_nov2001.pdf; Adalberth, K. (1997) Energy use during the life cycle of buildings: a method. Building and Environment: Elsevier, https://doi.org/10.1016/S0360-1323(96)00068-6; Chen, T. Y, et al. (2001) Analysis of embodied energy use in the residential building of Hong Kong. Energy: Elsevier. https://doi.org/10.1016/S0360-5442(01)00006-8 [18, 19, 20].

these construction materials, which include concrete, steel, wood, masonry, glass, and other critical supplies like cement, insulation, paint, and gypsum, remain part of a complex and often opaque global supply chain [16]. Figure 2.3 demonstrates the embodied energy for commonly used building materials. For structural materials, previous studies showed the steel frame was 1.13 times greater than the wood frame, and the embodied energy of the concrete frame was 1.05 times greater than that of the wood frame [17]. In this chapter, we cannot cover all material; let's delve into the four most commonly used materials.

Let's start with *cement*; it is a fundamental ingredient in making concrete. It is also a significant pollutant, responsible for approximately 5% of global emissions [21, 22]. Figure 2.4 paints a picture of the dynamic supply chain of cement. In 2023, China led cement production with 2,100 million tons, followed by India, Vietnam, the United States, and Turkey producing 410, 110, 91, and 79 million tons, respectively [23]. On the flip side, the United States topped cement importation, spending $2,700 million, succeeded by the Philippines ($638.1 million), China ($600.0 million), the United

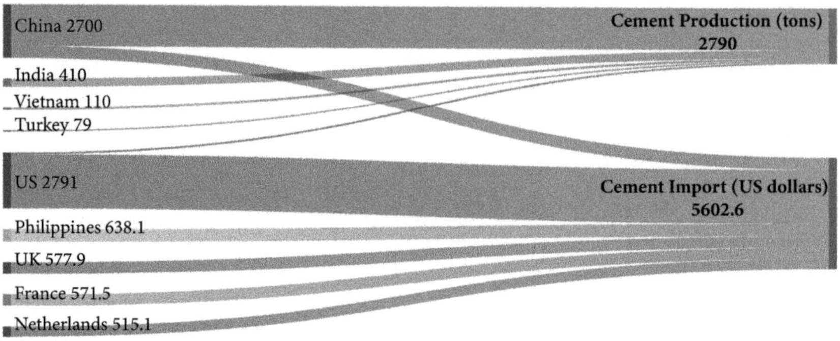

Figure 2.4 Cement production and import in a global context.

Source: data from Statista (2024). Major countries in worldwide cement production in 2023 and Statista (2024). Import value of cement worldwide in 2022, by select country. https://www.statista.com [23, 24].

Kingdom ($577.9 million), France ($571.5 million), and the Netherlands ($515.1 million) [24].

The second material is *gypsum*. It is an ancient building material that is plentiful in the Earth's crust and mined for use in products like cement and plaster. The international gypsum trade was vibrant in 2022, with trading exceeding $1.84 billion among 135 countries [25]. Just like cement, the supply chain of gypsum is very much globalized, as shown in Figure 2.5. Thailand dominated this export market, contributing 14%, closely followed by Oman (11.2%) and Spain (11.1%), with other significant exporters including Germany (11%) and Turkey (8.78%) [25]. The largest importer was the United States, absorbing 10.3% of the international market. India and Nigeria were also major importers, with substantial shares of 8.95 and 8.4%, respectively. Other key markets for gypsum importers included Japan (6.9%), Indonesia (4.91%), and Belgium (4.04%) [25], demonstrating the extensive global network of construction material trade and the strategic economic importance of gypsum.

Glass stands as a third material of note, prized for its transparency and durability, serving both practicality and aesthetics. Predominantly made of silica, its uses span from everyday items like bottles and eyewear to artistic expressions in stained glass windows and ornate tiles. In 2022, the top silica sand exporters, the primary raw material for glass, were the United States, Australia, Malaysia, Belgium, and Germany. These raw materials travel globally, fuelling the creation of various glass products [26, 27]. The global glass industry is spearheaded by five countries—China, Germany, the United States, Japan, and Italy. These nations collectively pushed more than half of the total exports

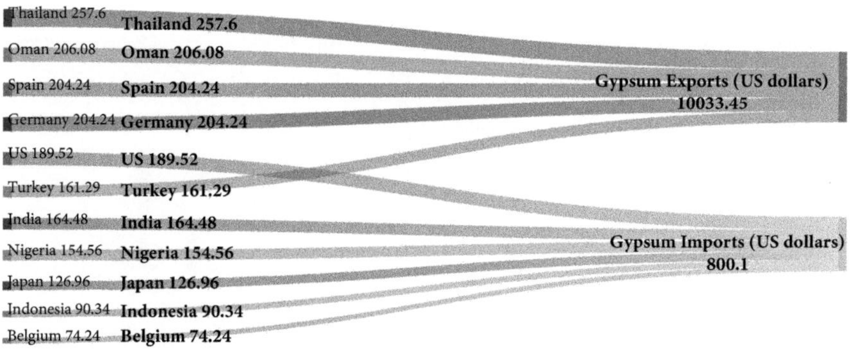

Figure 2.5 Gypsum import and export in a global context.

Source: data from Trend Economy. (2024). Gypsum; anhydrite; plasters. https://trendeconomy.com

in glass and glassware. Germany led with a 23% share in exports, with the U.S. closely behind at 21.6%, and Japan and China contributed 19.1% [26, 27]. Import patterns reveal China as the largest importer of flat glass (used in buildings), followed by Hong Kong SAR, Canada, Turkey, and India, reflecting a diverse demand for glass products across different regions [28]. Figure 2.6 demonstrates the complicated yet intricate global supply chain of glass products. Despite glass being 100% recyclable materials, the current global recycling rate for flat glass is very low, estimated to be only around 11% [29, 30]. This suggests that there is room for improvement in increasing the recycling rates for flat glass products to reduce waste generation and promote sustainability in the glass industry.

The last example is *aggregates*. Construction aggregates, including gravel, crushed stone, and sand extracted from pits and quarries, command the largest global construction materials market share. Concrete bricks and cement follow behind. As reported in 2023, the industry sand (fine aggregate) and gravel (coarse aggregate) producers are the United States [14], followed by China, Italy, Turkey, and France [31]. At the same time, the major importers are Canada (95%), Mexico (2%), and other countries (3%) [32].

The global supply chains for construction materials like cement, gypsum, glass, and aggregates are complex and expansive, involving multiple countries in production and trade. This complexity makes tracking and measuring the embodied carbon challenging due to varied sources, extensive transportation, and disparate regulatory environments. The intricate networks highlight the difficulty in ensuring transparency and accuracy in carbon accounting across different stages and regions.

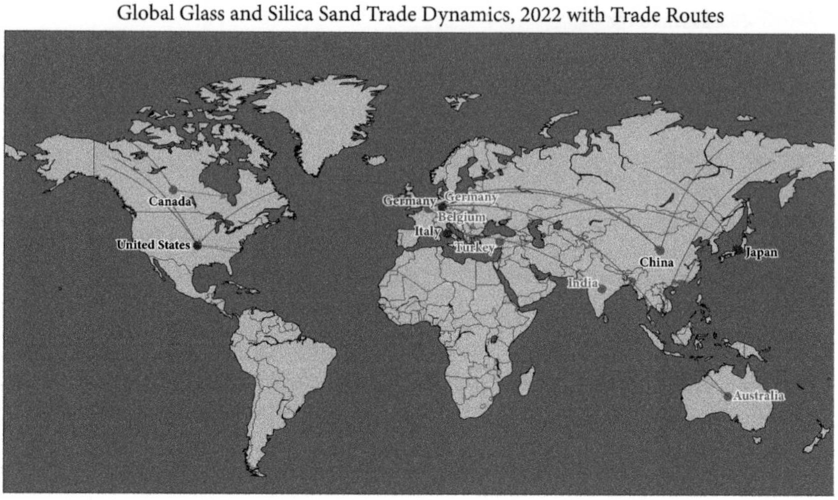

Figure 2.6 Global glass product supply chain.

Source: data from OEC (2024). Glass. https://oec.world/en/profile/sitc/glass-44664; Workman, D. (2024). Top glass and glassware exports by country. https://www.worldstopexports.com/top-glass-and-glassware-exports-by-country/; Indexbox, (2024). *The World's Best Import Market for Flat Glass*; Recovery, (2024). *Glass recycling—Current market trends.* https://www.recovery-worldwide.com/en/artikel/glass-recycling-current-market-trends-3248774.html [26, 27, 28].

2.3.2 How are different buildings made? Life-cycle stage A2–A5

Material production occurs during the life-cycle A-stages and is impacted by geographic location and technology accessibility. The prevailing understanding in sustainable construction posits that locally sourced materials and methods are typically less embodied energy-intensive and more carbon-conservative [33]. Indeed, a study on a pilot passive building in Iran employing traditional materials and methods demonstrated superior performance, with a 43 and 48% reduction in embodied energy and carbon, respectively [34].

Geographic and climatic uniqueness, along with divergent practices in raw material procurement and construction method, contribute to the disparities observed in embodied energy metrics across different countries and regions for life-cycle stage A [35]. For example, the creation of metal studs used in constructing partition walls varies widely depending on whether they are produced in the United States, Europe, or Asia due to differences in manufacturing protocols, transportation methods, and the distances materials must travel. These factors—extending to include the type of energy utilized in transportation, the distance covered, and even labour practices—collectively influence the embodied energy associated with building materials and assemblies. The

International Standard ISO 14040 underscores the necessity for data to accurately represent local specific data in assessing the life-cycle impacts of a building. This local-based data representativeness is vital to grasping the nuanced differences between embodied carbon and energy.

To demonstrate the impact of location and climatic conditions on embodied carbon, the author conducted an analysis of the 'Cradle to Gate' life-cycle embodied carbon (EC1) for eight prototype buildings based on the International Energy Conservation Code (IECC), representing Climate Zones 1 through 8 (refer to Figure 2.7 for climate zone definitions in the U.S. and Figure 2.8 for climate zone prototypes). The prototype is a single-family residential home plan (185 m^2). Each home shares a similar layout, with differences in foundation type, insulation, roof structure, and finishes based on typical, climate-specific practices.

The results (refer to Figure 2.8) indicate that colder climate zones exhibit higher embodied carbon due to the increased insulation requirements. For instance, in the hot and very hot Climate Zone 1, the required roof insulation is R-20 without rigid insulation, while in the cold and very cold Climate Zone 7, the required roof insulation is R-35 with rigid insulation. R-value measures a material's resistance to heat flow through conduction. The prototype building in Climate Zone 7 has nearly 50% more embodied carbon than that in Climate Zone 1. Additionally, the results show that in all climate zones, the top three contributors to embodied carbon are external walls, roofs, and foundations.

The study in one country reveals significant variation in embodied carbon, suggesting that an even greater range of variation can be observed globally due

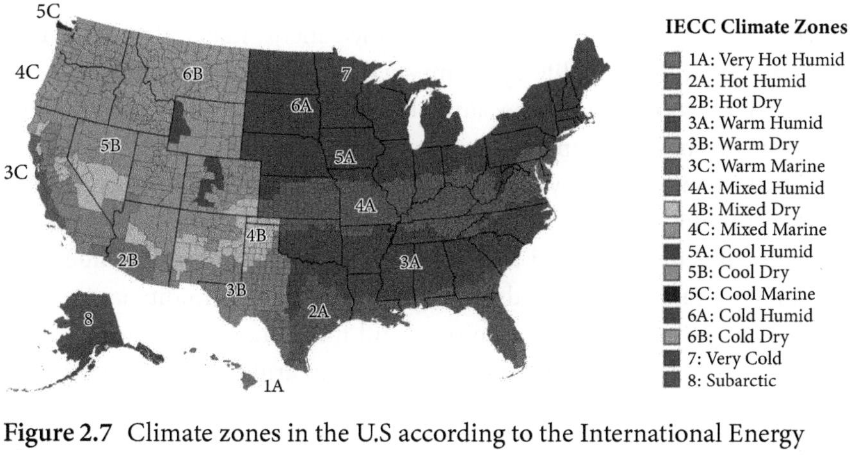

Figure 2.7 Climate zones in the U.S according to the International Energy Conservation Code definition.

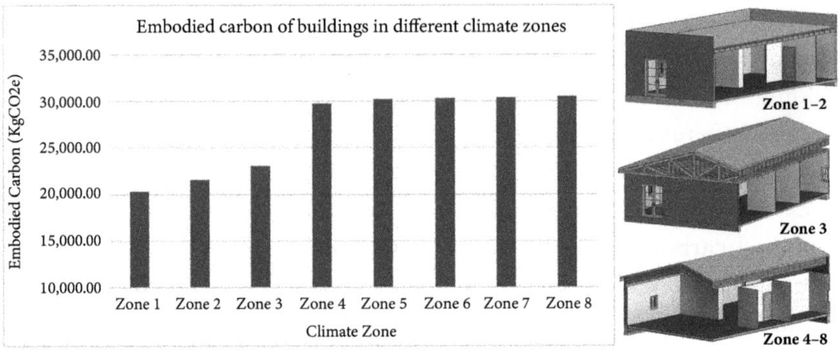

Figure 2.8 Analysis results. (left) Embodied carbon of prototype building in Climate Zones 1 to 8. (right) Prototypes buildings.

to differences in location, climatic conditions, construction types, and material choices. Therefore, it is crucial to fully understand localized needs and develop practical solutions tailored to specific regions, rather than applying a one-size-fits-all approach. Notably, conditions and solutions from the Global North should not be assumed to be directly applicable to the Global South without appropriate modifications.

2.3.3 Maintenance requirement: Life-cycle stage B

Material durability and climatic conditions closely determine maintenance requirements during life-cycle stage B. The durability and longevity of building materials are critical factors when calculating the embodied carbon for most existing buildings [36]. Higher embodied energy materials, such as brick, may have a lower overall life-cycle impact than lower embodied energy products, such as wood cladding because brick masonry lasts longer and requires fewer repairs and maintenance.

Climatic conditions play a crucial role in shaping the longevity and performance of buildings and materials. Factors such as temperature fluctuations, humidity, precipitation, and exposure to sunlight significantly influence the durability of building materials. In regions with high humidity and frequent rainfall, materials resistant to moisture and mold are essential for maintaining structural integrity over time. Conversely, in arid climates, materials that can withstand thermal cycling and intense UV exposure are preferable to minimize degradation. Additionally, climatic conditions inform the design of buildings, where features such as insulation, ventilation, window placement, and shading are meticulously planned to cater to local weather patterns, enhancing the building's resilience and energy efficiency. In extreme climates, buildings must be specially designed to cope with the conditions; for example, incorporating features like thick walls in hot environments to maintain interior coolness or steeply pitched roofs in snowy regions to prevent accumulation.

A common approach in Global North developed countries to reducing energy use under different climate conditions is to control the construction method by advancing the building code, which specifies insulation and material usage. Those requirements typically reduce operational energy; however, they can increase embodied carbon.

2.4 How does climate change affect embodied carbon?

It is predicted that in the next decades, more than half of the world's building additions will be concentrated in hot climate regions, primarily in Global South countries, which are experiencing a significant rise in space cooling demand. These regions often lack robust building energy codes and, crucially, the means to enforce them. Operationally, there is concern that advancements in energy efficiency made in colder climates could be negated by less efficient construction practices in these warmer regions. As cooling needs grow, there is also the looming issue of increased embodied carbon emissions, particularly if the pace of renewable energy adoption does not keep up with electricity demand. On the topic of embodied carbon, the crux of the issue lies in areas where construction is booming but the use of low-carbon materials is not prevalent. Such scenarios may result in locking in elevated levels of embodied carbon emissions over the long term. This is a pressing issue, as the choices made today in terms of construction materials and practices will have long-lasting impacts on our carbon footprint [37]. So, in section 2.5, we will examine what the global community has done to regulate, monitor, and reduce embodied carbon.

2.5 What is an environmental product label?

2.5.1 Market demand

Calculating the cumulation of embodied carbon from products and materials is essential to assessing a building's whole life cycle. Therefore, it is imperative to examine how we assess products' carbon footprints in the building and construction sector. As we strive for greater accountability in the construction industry's sustainability efforts, the demand for transparency around the environmental impact of products and services has surged. This push for clarity is powered by a coalition of stakeholders—from industry purchasers and individual consumers to advocacy groups and governmental bodies—all calling for a more transparent reporting of environmental performance. Here, we unpack the mechanisms behind product-level reporting and the history and types of environmental labels and declarations.

The advent and utilization of environmental labels and declarations serve as tools for this environmental communication, rooted in a rich history exemplified by pioneering initiatives. Notable among these are Germany's Demeter label, which was established in 1928 to denote products stemming from biodynamic agriculture [38], and the Blue Angel label from 1978, representing the vanguard of ecolabels with its comprehensive criteria and life-cycle perspective [39]. The Blue Angel label's introduction sparked a worldwide proliferation of environmental labelling and declarations.

In response to this widespread interest, the European Union has been refining the Product Environmental Footprint (PEF) methodology since 2011, aiming for significant ramifications upon its formal adoption into EU legislation. These shifts and the growing body of environmental labels and declarations underscore the need for a standardized approach. To this end, the International Organization for Standardization (ISO) launched the ISO 14020 series of standards, which include ISO 14020:2000, 14025:2006, 14021:2016, and 14024:2018, establishing foundational principles for communicating environmental performance. These standards have become cornerstones in the edifice of building sector credibility, ensuring the environmental claims made are as sturdy and transparent as the structures we inhabit.

2.5.2 Types of labels

In the pursuit of environmental integrity, the ISO 14020 series emerges as a vital framework, categorizing methodologies for representing the environmental performance of products and services [40]. These are meticulously documented within the ISO standards and are stratified into three types [41]. Each type upholds a unique set of criteria and goals, ensuring the dependability and clarity of environmental claims, as shown in Box 2.2.

Box 2.2: Three types of ecolabels

- Type I methodology involves a multicriteria-based label guided by an external audit to certify environmental assertions [40]. It stands as one of the first stringent of the standards, requiring a comprehensive LCA and a set of predefined criteria.
- Type II is a self-declared environmental claim, an organization's own proclamation of environmental performance, not reliant on external verification [41]. This type tends to focus on specific attributes such as recycled content, energy efficiency, or biodegradability.
- Type III declarations represent an apex of transparency, presenting full disclosure on the life-cycle impact [42]. These detailed assessments require collecting and evaluating life-cycle inventory data by an independent third party, resulting in a declaration revealing a product's comprehensive environmental footprint.

Organizations may employ one or more of these methodologies to align with regulatory requirements or stakeholder expectations across various regions. This strategic flexibility includes the development of hybrid labels that blend attributes from different standard types to meet the demands of a more complex sustainability landscape. Such is the case with the ISO 14067 standard on the carbon footprint of products (CFP), reflecting the specific information needs of today's market [42].

Type I environmental labels

Type I environmental labels are instantly recognizable, adorning a vast range of consumer products from the food in our baskets to the clothes in our wardrobes and the furniture that adorns our living spaces. These labels serve as a compass for the eco-conscious consumer, offering insights into the environmental attributes of products and services. The three earliest established ecolabels are Blue Angel, Nordic Swan, and EU Ecolabel. The Blue Angel, introduced in 1978, is recognized as the first Type I label, setting a precedent for environmental certification [39]. The Nordic Swan, established in 1989, exemplifies voluntary ecolabelling in Scandinavian countries [43]. The EU Ecolabel, launched by the European Commission in 1992, stands out as the only pan-European Type I ecolabel in line with ISO 14024 [44].

The philosophy underpinning Type I labels is founded on precision and verifiability, with a focus on substantial environmental issues to avoid any misdirection. The formulation of these criteria is not an isolated process; it is a synergy of diverse stakeholder input underpinned by thorough scientific research. Furthermore, the governance of these labels is entrusted to impartial entities, guaranteeing transparency and objectivity in all procedural and methodological aspects.

Type II environmental labels

In contrast to the multiattribute Type I labels, Type II labels are single-attribute, primarily developed by manufacturers to highlight one particular aspect of a product's environmental impact. These labels represent the manufacturer's self-declared environmental claims, which, unlike third-party audited certifications, rely on the integrity of the producer rather than independent validation. The creation of the ISO 14021 standard came as a direct response to the burgeoning number of environmental claims, establishing a need for a system to verify their authenticity. Claims under Type II labels typically focus on specific environmental benefits like recycled content, recyclability, biodegradability, and energy efficiency.

Transparency stands as the guiding principle for Type II labels, tasking manufacturers with the responsibility to provide comprehensive information about their products' environmental attributes. This level of detail is crucial to facilitate independent verification, despite these claims not being subject to the stringent criteria that third-party certifications require. Hence, the onus is on consumers to appraise the environmental merits of these products, relying on the trustworthiness of the information presented by the manufacturers.

Well-known examples of Type II labels include the Energy Star Label, which signals energy efficiency in appliances, and the WaterSense label, denoting water conservation in products. Another instance is the Dolphin-safe label, which assures consumers of adherence to practices aimed at reducing dolphin harm in fishing operations.

Type III environmental labels

Within ecolabelling frameworks, Type III labels stand out by offering a holistic view of a product's environmental impact, contrasting with the partial insights provided by Type I labels. Unlike Type II labels, which are limited to by manufacturers' self-declared claims, Type III labels encapsulate a full life-cycle sustainability analysis, providing a transparent, exhaustive account of a product's environmental impact [45]. These labels are particularly relevant in the building and construction sector, where detailed environmental data are pivotal for informed decision-making.

Type III environmental labels represent an advanced ecolabelling system that encompasses a product's complete life cycle, from production to disposal. The process demands a rigorous assessment by an independent party, responsible for gathering life-cycle inventory data and measuring it against quantifiable environmental parameters, such as emissions and energy consumption. The Environmental Product Declaration (EPD), a prevalent Type III label, epitomizes this approach, offering an official declaration based on standardized product category rules and LCA techniques [46, 47].

Beyond internationally recognized EPDs, various regions have developed their own Type III labels. For instance, the Global Environmental Declarations Network (GEDnet) facilitates global dialogue and knowledge sharing among creators of Type III declarations [48]. Meanwhile, the Eco-leaf label in Japan [49], Thailand's Carbon Footprint Reduction Label, Korea's Carbon Footprint Labelling, and the global Ecogloballabel NSO-NAP 3 Standard are examples of Type III labels tailored to regional standards [50], reflecting a global commitment to sustainability through comprehensive environmental labelling. But how effective are those labels? Let's take a look at EPD as an example in section 2.6. It is one of the most influential Type III labels in the built environment sector that aims to enhance the transparency of environmental claims, fostering a marketplace where sustainable choices are accessible and data-driven [51].

2.6 The complex realities behind ecolabels

2.6.1 Factors behind EPD expansion

The burgeoning presence of ecolabel programmes, particularly EPD, can be traced to three principal drivers: regulatory impetus, advancements in the digital automation of certification processes, and consumer demand. Regulatory demand stands out as the primary propellant of this expansion. In Germany, for example, Green Public Procurement (GPP) policies require public buildings to meet certain embodied carbon thresholds, verified through LCAs. GPP's leverage is significant, with German authorities wielding an annual purchasing power of roughly €300 billion, poised to reduce environmental impact and champion eco-friendly innovations. According to its website, GPP is described as:

> Each year, German contracting authorities purchase around €300 billion worth of products ranging from pencils to city buses. This massive purchasing power can potentially be used to reduce environmental pollution, improve environmental product and service range, and promote more targeted market launches of innovative eco-friendly products. [52]

In Norway, GPP strategies have been deliberately utilized to bolster EPD adoption, while French regulations have significantly encouraged their creation. Similarly, in the United States, 'Buy Clean' and 'Low carbon concrete' initiatives at the state and county levels have noticeably catalyzed the proliferation of EPDs and other ecolabel programmes [2]. The concerted backing of national governments in the Global North—including the U.S., U.K., France, Germany, and the Netherlands—for the production of generic datasets and the upkeep of national databases underscores a substantial commitment to ecolabels (e.g. EPD) dissemination, consequently to promote the sustainable practice.

This trend is further reinforced by financial support mechanisms, such as the United States' Inflation Reduction Act of 2022, which earmarked $250 million to support EPD generation among companies [53]. Additional incentives include Enterprise Ireland funding for EPD production [54] and France's reduced EPD registration fees—all contributing to the uptick in EPD development [55]. The influence of green building certifications, like the LEED EPD credit in the U.S., also contributes significantly to increasing EPD counts, as demonstrated by the surge within the UL Environment EPD programme [55]. Under LEED v4.1, the Material and Resources (MR) credit category mandates that new construction and major renovations prioritize waste reduction and recycling, including provisions for recyclable material storage and collection during building occupancy [56].

The second propellant for the rise in EPD adoption is the streamlining of production via automated EPD tools, as seen in the high volume of EPDs generated using such tools in France, Germany, and the U.S. These tools, such as France's Échanges de Données Environnementales Configurées (EDEC), allow for efficient EPD storage and integration into building assessment frameworks without separate registrations. They utilize preverified data and models that conform to established product category rules (PCRs), exemplified by

Table 2.1 EPD tools

Name of EPD tool	Year	Country
2016 BRE Lina EPD Tool launch	2016	UK
NORSUS web-based EPD Tool (2012)	2012	Norway
NORSUS first preverified EPD generator	2013	Norway
EPD.digi	2022	Norway
Industry EPD tools launch	2011	France
inies webservice	2013	France
Extension of ILCD for EPD published	2013	Germany
BRE Lina EPD Tool launch	2016	UK

tools like the OneClick LCA EPD generator. Table 2.1 lists some EPD tools in the Global North.

The digitization of ecolabels emerges as a critical driver of their widespread adoption and effectiveness, streamlining ease of access and integration into various analytical tools and assessments [57]. This advancement is exemplified by successful European initiatives, including France's inies database [58] and Germany's oekobaudat database [59]. Norway is at the forefront of digital EPD innovation as well, leveraging EN ISO 22057:2021 standards to integrate EPD data with Building Information Modeling (BIM) systems, thereby enhancing collaborative and informed building design. Similarly, the Embodied Carbon in Construction Calculator (EC3) tool in the United States has revolutionized EPD accessibility, providing a comprehensive platform for industry professionals to assess and compare the environmental impacts of construction materials with unprecedented ease. Transitioning EPDs to digital platforms has significantly improved the accessibility, sharing, and analytical evaluation of environmental data. This evolution supports the broader digital transformation of industries, refining decision-making processes and strengthening the pursuit of sustainability across sectors. The readily available digital format of EPD data also facilitates robust comparisons of environmental performance across products, bolstering sustainability initiatives and promoting environmentally responsible choices.

The third factor is consumer demand. Consumers play a crucial role in driving the demand for environmentally friendly products. Studies have shown that consumers increasingly seek information on products' environmental impacts and companies' corporate social responsibility performance [60]. Consumer behaviour towards green products is influenced by various factors, including environmental consciousness, attitudes, values, and knowledge [61].

This knowledge not only refers to the general concept of sustainability but also to specific information about the products they are considering purchasing [62]. Hence, information on personal gain from choosing green products is critical. The gap between consumer purchasing intentions and actual behaviour can be narrowed through ecolabelling awareness and education initiatives, which aim to align consumer decision-making processes with their self-declared commitment to sustainability [63]. EPDs serve as verified environmental profiles that communicate quantitative information about products to both manufacturers and distributors (business-to-business) and final consumers (business-to-consumer) [64]. Considering these factors together, it is evident that the synergistic effects of regulatory initiatives and technological innovation are not only enriching ecolabel creation and application but are also propelling the prominence of EPDs in the ecolabel marketplace. But EPD is not all rosy without any concerns. Can ecolabels like EPD lead to global sustainability? The answer remains unseen.

2.6.2 Global disparities in EPD production

With over 130,000 verified EPDs now registered, the international EPD system has experienced considerable growth [55]. This expansion has facilitated the emergence of numerous EPD programmes worldwide, with significant participation from European countries. However, notable disparities in global distribution emphasize the importance of inclusivity and fairness in the ecolabelling domain.

In the building and construction sector, EPD production shows a marked imbalance, with a clear bias towards the Global North. Key players such as France's Fiches de Déclaration Environnementale et Sanitaire (FDES), Sweden's International EPD, Norway's EPD Norge, Germany's Institut Bauen und Umwelt (IBU), and UL Environment in the United States are leading the way [55]. There are around 16,000 EPDs from the built sector worldwide, and FDES accounts for 4,585 EPDs. Following closely are International EPD with 2,945, EPD Norge with 1,977, IBU with 1,302, and UL with 1,279 [55]. Together, these entities account for over two-thirds of the global tally, highlighting a significant concentration of EPDs within a few Global North countries.

In stark contrast, the Global South exhibits a notable deficiency in EPD production and regulation. For instance, EPD China, founded in 2020, had only two published EPDs by March 2024 [65]. Moreover, a review of the International EPD library indicates severe under-representation in the Global

South, with only fifty EPDs linked to China, thirty-five to Africa, and thirty-six to Latin America [66]. The reason for the shortage is illustrated in Box 2.3.

Box 2.3: Reason for shortage of EPD in Global South

The lag in EPD or other ecolabel development in Global South countries is not due to a lack of willingness to adopt environmentally friendly products. Instead, financial constraints and technical gaps impede the development of EPDs. According to the author's personal experience, hiring a professional to conduct a LCA to generate EPD documents and have them certified by a registered entity costs over US$30,000 per product, a figure corroborated by other EPD providers such as OneClick LCA. Such high expenses are unaffordable for most developing countries. Moreover, there are very few certified life-cycle modellers and EPD verifiers in countries less familiar with those rules.

The question remains: should the Global North-originated rules, procedures, and standards be applied to the Global South without any modification? This profound geographical skewness influences the EPD market and the shaping of regulatory landscapes, raising significant questions about the equity and validity of the current ecolabelling system. This imbalance calls into question the sustainability practices within the global built environment. There is a clear need for a critical assessment, supported by empirical data and scientific rigour, to establish an ecolabelling framework that is both inclusive and representative, ensuring equitable impact and integration of diverse global perspectives.

2.6.3 Ecolabel credibility challenge

Research has consistently shown that ecolabels can foster environmentally friendly behaviours, revealing that consumers are often willing to pay a premium for products adorned with ecolabels [67]. Indeed, some argue that even the phenomenon of greenwashing—misleading claims about a product's environmental friendliness—can alter consumer behaviour positively [68]. Ideally, ecolabels, such as EPDs, should equip professionals and consumers with reliable information to make environmentally sound decisions.

However, the integrity of EPDs is sometimes compromised. A notable example is some manufacturers' use of loopholes within EPD standards to obscure the potential toxicity of their products, a practice that could pose risks to both human health and the environment. Forbo Flooring, for instance, has pointed out that these gaps in flooring EPDs, while not immediately obvious, have significant repercussions [69]. These issues often originate from manipulated PCRs, the technical underpinnings of EPDs, leading to the exclusion of crucial ecotoxicity and human toxicity indicators and thus, diminishing the declaration's efficacy [69]. Defenders of such practices claim a lack of comprehensive data to fully represent these impacts in an LCA, arguing that emissions testing should be the sole measure of a product's health effects. This ongoing contention has allowed products like PVC and rubber flooring to be marketed as environmentally sound under a façade of green labelling, distorting their true ecological footprint.

Another example was pointed out by the Sierra Club: under the current ISO 14025 rules, manufacturers are not required to publish their products' impact on human health and ecological toxicity in EPD. Toxins can be released into the environment during a product's extraction and manufacturing phase (A1–A3), and also leach into the environment during a product's end-of-life and disposal phase (C1–C4). Within the context of the global supply chain, life cycle stages A1–A3 (raw material extraction, manufacturing) and C1–C4 (demolition, disposal, and waste process) mainly occur in the Global South, and consumption and demand originate in the Global North. Such shifting of the human health burden and ecological system burden from the Global North to Global South was not captured by many ecological systems, especially EPDs [70].

While ecolabels are meant to guide consumers towards more sustainable choices, this guidance becomes unreliable if the labels are not all-encompassing and trustworthy. Experts are raising alarms about industries' strategic use of weaker labels to circumvent stricter regulations [71]. Additionally, the market is becoming cluttered with misleading labels that resemble credible certifications, causing confusion among consumers and potentially leading them to make choices that are not truly environmentally friendly.

The crux of the matter is that for ecolabels to fulfil their purpose effectively, there must be stringent oversight to prevent the dilution of their credibility. A robust, transparent, and honest ecolabelling system is essential to ensure that environmental claims are genuine, leading to better consumer choices and industry practices. Given that the primary aim of most ecolabels in the building and construction sectors is to reduce embodied carbon, a more comprehensive explanation and comparison of embodied carbon policies and programmes worldwide is provided in Chapter 7.

2.7 Conclusion

Embodied carbon in the built environment sector represents a pivotal front in the global effort to mitigate climate change. Chapter 2 delves into the complexities of defining and managing embodied carbon, highlighting the critical role of LCA methodologies in quantifying the environmental impact of construction materials from extraction through to disposal. The discussion reveals a landscape marked by evolving regulatory frameworks, advancing technological tools, and a shifting market demand towards sustainability and transparency.

Chapter 2 also outlines a nuanced understanding of environmental product labels and declarations, underscoring the importance of accurate, verifiable, and comprehensive environmental reporting. It showcases the diversity of labelling standards—from Type I to Type III labels—each serving distinct purposes but collectively contributing to a more informed and environmentally conscious market. The EPDs emerge as a testament to the built environment sector's growing commitment to sustainability, albeit with an observed imbalance favouring the Global North, indicating a need for more inclusive and equitable global sustainability practices.

This imbalance affects the ecolabel market and shapes regulatory landscapes, raising concerns about the equity and validity of the ecolabelling system. The skewed distribution suggests a need for a more inclusive and representative ecolabelling framework that integrates diverse global perspectives, ensuring that sustainable practices and regulations in the built environment are equitably distributed and applied worldwide.

Chapter 3 will connect embodied carbon to climate change and environmental justice, highlighting how the imbalance observed in EPD development is actually reflected more broadly in current sustainable practice globally.

References

1. Adams M, Burrows V, Richardson S, Drinkwater J, Gamboa C, Collin C, et al. Bringing embodied carbon upfront: Coordinated action for the building and construction sector to tackle embodied carbon [Internet]. 2019 [Accessed 9 Sep 2024]. Available

from: https://worldgbc.s3.eu-west-2.amazonaws.com/wp-content/uploads/2022/09/22123951/WorldGBC_Bringing_Embodied_Carbon_Upfront.pdf
2. Hu M, Esram NW. The status of embodied carbon in building practice and research in the United States: a systematic investigation. Sustainability. 2021 Nov 23; 13(23): 12961.
3. Müller LJ, Kätelhön A, Bachmann M, Zimmermann A, Sternberg A, Bardow A. A guideline for life cycle assessment of carbon capture and utilization. Front Energy Res. 2020 Feb 14; 8: 15.
4. ISO. ISO 14044:2006/Amd 2:2020 Environmental management—Life cycle assessment—requirements and guidelines—Amendment 2 [Internet]. 2020 [Accessed 2 Oct 2022]. Available from: https://www.iso.org/standard/76122.html
5. Liu H, Li J, Sun Y, Wang Y, Zhao H. Estimation method of carbon emissions in the embodied phase of low carbon building. Advances in Civil Engineering. 2020 Jan; 2020(1): 8853536.
6. Xie X, Cai W, Jiang Y, Zeng W. Carbon footprints and embodied carbon flows analysis for China's eight regions: a new perspective for mitigation solutions. Sustainability. 2015 Jul 28; 7(8): 10098–114.
7. Blay-Armah A, Bahadori-Jahromi A, Mylona A, Barthorpe M, Ferri M. An evaluation of the impact of databases on end-of-life embodied carbon estimation. Sustainability. 2022 Feb 17; 14(4): 2307.
8. Varun, Sharma A, Shree V, Nautiyal H. Life cycle environmental assessment of an educational building in Northern India: a case study. Sustainable Cities and Society. 2012 Oct; 4: 22–8.
9. Chastas P, Theodosiou T, Kontoleon KJ, Bikas D. Normalising and assessing carbon emissions in the building sector: a review on the embodied CO_2 emissions of residential buildings. Building and Environment. 2018 Feb; 130: 212–26.
10. Kofoworola OF, Gheewala SH. Environmental life cycle assessment of a commercial office building in Thailand. International Journal of Life Cycle Assessment. 2008; 13(6): 498.
11. Blengini GA. Life cycle of buildings, demolition and recycling potential: a case study in Turin, Italy. Building and Environment. 2009 Feb; 44(2): 319–30.
12. Global Construction Perspectives. Top 3 Construction Markets Globally Are in Asia. US is 4th—Global Construction Perspectives [Internet]. 2024. Available from: https://www.prnewswire.com/news-releases/top-3-construction-markets-globally-are-in-asia-us-is-4th—global-construction-perspectives-302041155.html
13. Economic Consulting Team. Global Construction Futures [Internet]. [Accessed 12 Apr 2024]. Available from: https://www.oxfordeconomics.com/resource/global-construction-futures/
14. Precedence Research. Construction Materials Market [Internet]. 2023 Mar [Accessed 11 Apr 2024]. Report No. 2716. Available from: https://www.precedenceresearch.com/construction-materials-market
15. U.S. Department of Transportation. Bipartisan Infrastructure Law (BIL)/Infrastructure Investment and Jobs Act (IIJA) [Internet]. [Accessed 12 Apr 2024]. Available from: https://www.phmsa.dot.gov/legislative-mandates/bipartisan-infrastructure-law-bil-infrastructure-investment-and-jobs-act-iija

16. Jang HJ, Ahn YH, Tae SH. Proposal of major environmental impact categories of construction materials based on life cycle impact assessments. Materials. 2022 Jul 20; 15(14): 5047.
17. Cole RJ, Kernan PC. Life-cycle energy use in office buildings. Building and Environment. 1996; 31(4): 307–17.
18. Adalberth K. Energy use during the life cycle of buildings: a method. Building and Environment. 1997; 32(4): 317–20.
19. Alcorn A. Embodied energy and CO coefficients for New Zealand building materials [Internet]. Wellington, New Zealand: Centre for Building Performance Research, Victoria University of Wellington, 2001 Nov [Accessed 24 Jul 2024]. Available from: https://www.wgtn.ac.nz/architecture/centres/cbpr/resources/pdfs/ee-co2_report_nov2001.pdf
20. Chen T, Burnett J, Chau CK. Analysis of embodied energy use in the residential building of Hong Kong. Energy. 2001; 26(4): 323–40.
21. Cheng D, Reiner DM, Yang F, Cui C, Meng J, Shan Y, et al. Projecting future carbon emissions from cement production in developing countries. Nat Commun. 2023 Dec 11; 14(1): 8213.
22. Xi F, Davis SJ, Ciais P, Crawford-Brown D, Guan D, Pade C, et al. Substantial global carbon uptake by cement carbonation. Nature Geosci. 2016 Dec; 9(12): 880–3.
23. statista. Major countries in worldwide cement production in 2023 [Internet]. [Accessed 14 Apr 2024]. Available from https://www.statista.com/statistics/267364/world-cement-production-by-country/#:~:text=China%20produces%20the%20most%20cement,half%20of%20the%20world's%20cement
24. statista. Import value of cement worldwide in 2022, by select country [Internet]. [Accessed 14 Apr 2024]. Available from: https://www.statista.com/statistics/1086042/countries-cement-import-value-globally/
25. Trend Economy. Gypsum; anhydrite; plasters | Imports and exports | 2022 [Internet]. [Accessed 14 Apr 2024]. Available from https://trendeconomy.com/data/commodity_h2/2520#:~:text=In%202022%2C%20the%20world%20imports,trade%20statistics%20of%20135%20countries
26. OEC. Glass [Internet]. [Accessed 12 Apr 2024]. Available from https://oec.world/en/profile/sitc/glass-44664?redirect=true
27. Workman D. Top glass and glassware exports by country [Internet]. [Accessed 12 Apr 2024]. Available from https://www.worldstopexports.com/top-glass-and-glassware-exports-by-country/
28. Indexbox. The world's best impor market for flat glass [Internet]. 2023 [Accessed 14 Apr 2024]. Available from https://www.indexbox.io/blog/world-worlds-best-import-markets-for-flat-glass/#:~:text=China,USD%203.1%20billion%20in%202022
29. Glass recycling—current market trends [Internet]. [cited 2024 Apr 14]. Available from https://www.recovery-worldwide.com/en/artikel/glass-recycling-current-market-trends-3248774.html
30. Geboes E, Galle W, De Temmerman N. Value network mapping applied to the flat glass sector: constraints and leverage points for circularity. IOP Conf Ser: Earth Environ Sci. 2022 Dec 1; 1122(1): 012033.

31. statista. Major countries in industrial sand and gravel production worldwide in 2023 [Internet]. [Accessed 14 Apr 2024]. Available from https://www.statista.com/statistics/376665/industrial-sand-and-gravel-production-by-top-countries/
32. Distribution of U.S. industrial sand and gravel imports between 2018 and 2021 [Internet]. [Accessed 14 Apr 2024]. Available from https://www.statista.com/statistics/376660/industrial-sand-and-gravel-import-sources-in-the-us/
33. Aktürk G, Fluck H. Vernacular heritage as a response to climate: lessons for future climate resilience from Rize, Turkey. Land. 2022 Feb 11; 11(2): 276.
34. Pakdel A, Ayatollahi H, Sattary S. Embodied energy and CO2 emissions of life cycle assessment (LCA) in the traditional and contemporary Iranian construction systems. Journal of Building Engineering. 2021 Jul; 39: 102310.
35. Dixit MK, Fernández-Solís JL, Lavy S, Culp CH. Identification of parameters for embodied energy measurement: A literature review. Energy and Buildings. 2010; 42(8): 1238–47.
36. Hu M. Building impact assessment—a combined life cycle assessment and multi-criteria decision analysis framework. Resources, Conservation and Recycling. 2019 Nov; 150: 104410–104410.
37. UN Environment Programme. 2023 Global Status Report for Buildings and Construction [Internet]. 2024 Mar [Accessed 9 Apr 2024]. Available from https://www.unep.org/resources/report/global-status-report-buildings-and-construction
38. History [Internet]. BIODYNAMIC. [Accessed 17 Jun 2024]. Available from https://demeter.net/about/history/
39. Blue Angel. Ecolabel with history [Internet]. [cited 18 Mar 2024]. Available from https://www.blauer-engel.de/en/blue-angel/our-label-environment/ecolabel-history
40. International Organization for Standardization. ISO 14020:2022 Environmental statements and programmes for products Principles and general requirements [Internet]. 2022 [cited 18 Mar 2024]. Available from https://www.iso.org/standard/79479.html
41. Skaar C. Communicating product life cycle performance through labels and declarations. In: Fet AM, editor. Business transitions: a path to sustainability [Internet]. Cham: Springer International Publishing; 2023 [cited 18 Mar 2024]. pp. 57–65. Available from https://link.springer.com/10.1007/978-3-031-22245-0_6
42. International Organization for Standardization. ISO 14067:2018 Greenhouse gases Carbon footprint of products Requirements and guidelines for quantification [Internet]. 2018 [Accessed 18 Mar]. Available from https://www.iso.org/standard/71206.html
43. Nordic Sawm Ecolabel. The Nordic Swan Ecolabel [Internet]. [Accessed 18 Mar 2024]. Available from https://www.nordic-swan-ecolabel.org/official-nordic-ecolabel/#:~:text=The%20Nordic%20Swan%20Ecolabel%20was,%2C%20Iceland%2C%20Norway%20and%20Sweden.
44. European Commission. About the EU Ecolabel [Internet]. [Accessed 18 Mar 2024]. Available from https://environment.ec.europa.eu/topics/circular-economy/eu-ecolabel/about-eu-ecolabel_en
45. Toniolo S, Mazzi A, Simonetto M, Zuliani F, Scipioni A. Mapping diffusion of Environmental Product Declarations released by European program operators. Sustainable Production and Consumption. 2019 Jan; 17: 85–94.

46. Sharma NK, Kushwaha GS. Eco-labels: a tool for green marketing or just a blind mirror for consumers. EGJ [Internet]. 2019 Feb 16 [Accessed 21 Mar 2024]; 1(42). Available from https://escholarship.org/uc/item/6k83s5mv
47. D'Souza C, Taghian M, Lamb P, Peretiatko R. Green decisions: demographics and consumer understanding of environmental labels. Int J Consumer Studies. 2007 Jul; 31(4): 371–6.
48. Global Type III Environmental Product Declarations Network (GEDnet) [Internet]. UIA Global Civil Society Database. [Accessed 21 Mar 2024]. Available from https://uia.org/s/or/en/1100010142
49. Japan Environmental Management Association for Industry. Eco-Leaf Environmental Declaration [Internet]. [Accessed 21 Mar 2024]. Available from https://www.ecoleaf-jemai.jp/application/data/pamph-e.pdf
50. Green/environmental certification or labelling schemes under the Eco Mark Recognition [Internet]. [Accessed 21 Mar 2024]. Available from https://www.ecogloballabel.org/download/GLOBAL_ECO_LABELS_LISTESI.pdf
51. Rochmadi T, Herdioso R, Isharyadi F, Rustianto B, Aziz A. Implementation of type III environmental declaration based on ISO 14025: a case study on an ideal and reality of the PROPER program implementation in Indonesia. IOP Conf Ser: Earth Environ Sci. 2022 Nov 1; 1108(1): 012023.
52. Umwelt Bundesamt. Green public procurement: beschaffung-info.de [Internet]. [Accessed 25 Mar 2024]. Available from https://www.umweltbundesamt.de/en/topics/economics-consumption/green-public-procurement
53. The White House. Building a clean energy economy: a guidebook to the Inflation Reduction Act's investments in clean energy and climate action [Internet]. 2023 Jan [Accessed 25 Mar 2024]. Available from https://www.whitehouse.gov/wp-content/uploads/2022/12/Inflation-Reduction-Act-Guidebook.pdf
54. IGBC I. How to get assistance from Enterprise Ireland & IDA to develop an Environmental Product Declaration [Internet]. 2020 [Accessed 8 Apr 2024]. Available from https://www.igbc.ie/how-to-get-assistance-from-enterprise-ireland-to-develop-an-environmental-product-declaration/#:~:text=Enterprise%20Ireland%20Grants&text=Specifically%20for%20developing%20EPD%20these,and%20improve%20your%20environmental%20performance
55. Anderson J. Reducing embodied carbon in the built environment: the role of Environmental Product Declarations. PhD thesis, The Open University, 2023.
56. Hughes E. How LEED v4.1 addresses embodied carbon [Internet]. 2019 [Accessed 25 Mar 2024]. Available from https://www.usgbc.org/articles/how-leed-v41-addresses-embodied-carbon
57. Rasmussen F, Andersen C, Wittchen A, Hansen R, Birgisdóttir H. Environmental Product Declarations of structural wood: a review of impacts and potential pitfalls for practice. Buildings. 2021 Aug 18; 11(8): 362.
58. inies. Product LCA [Internet]. [Accessed 8 Apr 2024]. Available from https://www.inies.fr/en/inies-for-building/product-lca/
59. ÖKOBAUDAT. ÖKOBAUDAT: Sustainable Construction Information Portal [Internet]. [Accessed 8 Apr 2024]. Available from https://www.oekobaudat.de/en.html

60. Young W, Hwang K, McDonald S, Oates CJ. Sustainable consumption: green consumer behaviour when purchasing products. Sustainable Development. 2010 Jan; 18(1): 20–31.
61. Laheri VK, Lim WM, Arya PK, Kumar S. A multidimensional lens of environmental consciousness: towards an environmentally conscious theory of planned behavior. JCM. 2024 Apr 26; 41(3): 281–97.
62. Amberg N, Fogarassy C. Green consumer behavior in the cosmetics market. Resources. 2019 Jul 30; 8(3): 137.
63. Calderon-Monge E, Redondo-Rodriguez RG, Ramírez-Hurtado JM. Narrowing the gap between consumer purchasing intention and behaviour through ecolabelling: a challenge for eco-entrepreneurism. BFJ. 2021 Oct 21; 123(10): 3293–308.
64. Ibáñez-Forés V, Pacheco-Blanco B, Capuz-Rizo SF, Bovea MD. Environmental Product Declarations: exploring their evolution and the factors affecting their demand in Europe. Journal of Cleaner Production. 2016; 116: 157–69.
65. EPD China. Published EPDs [Internet]. [Accessed 25 Mar 2024]. Available from http://www.epdchina.cn/publishedEPDs
66. EPD International. Search the EPD Library [Internet]. [Accessed 5 Mar 2024]. Available from https://www.environdec.com/library
67. Schwartz D, Loewenstein G, Agüero-Gaete L. Encouraging pro-environmental behaviour through green identity labelling. Nat Sustain. 2020 May 25; 3(9): 746–52.
68. Lopes JM, Gomes S, Trancoso T. The dark side of green marketing: how greenwashing affects circular consumption? Sustainability. 2023 Jul 28; 15(15): 11649.
69. Fobo Flooring. Health and hypocrisy [Internet]. [Accessed 2 Apr 2024]. Available from https://forbo.blob.core.windows.net/forbodocuments/22444/Health%20and%20Full%20Transparency.pdf
70. Sierra Club. The toxic loophole in Environmental Product Declarations (EPDs) [Internet]. [Accessed 8 Apr 2024]. Available from: https://www.sierraclub.org/sites/default/files/program/documents/EPD-Toxic%20Loophole%20One%20Page.pdf
71. Buckley R. Three reasons for eco-label failure. Nature. 2013 Aug 8; 500(7461): 151.

Chapter 3
Embodied carbon, environmental burden, and climate change

3.1 The need for addressing embodied carbon

3.1.1 Critical to climate action

While built environments fulfil our essential need for shelter and form the backbone of civilization, they are paradoxically contributing to the planet's ecological distress through intense resource consumption and environmental impact [1]. The building industry is a colossal consumer of resources, devouring 50% of the raw materials extracted globally and accounting for 38% of greenhouse gas emissions—a substantial environmental footprint. The dire implications of this were underscored in the 2023 Global Status Report for Buildings and Construction by the United Nations Environment Programme (UNEP), which starkly reminded us that the trajectory we are on is misaligned with the objectives of the Paris Agreement. Despite the dip in emissions in 2020 and 2021, an anomaly caused by the COVID-19 pandemic, there has been a worrying resurgence, shattering pre-pandemic peaks [2].

Alarmingly, the construction sector's energy demands and emissions have not only rebounded but have done so with disquieting speed, emphasizing the urgent need for climate action. The UNEP's report is a clarion call: By 2050, about 60% of the infrastructure required has yet to be constructed, and 20% of the existing buildings necessitate zero-carbon-ready renovations by 2030 [3]. At the 28th Conference of the Parties (COP28), an initiative named the Buildings Breakthrough was born, where twenty-seven nations pledged to mainstream near-zero emission and resilient buildings by the decade's end. Those countries are Armenia, Austria, Canada, China, Côte d'Ivoire, Egypt, Ethiopia, Finland, France, Germany, Guinea-Bissau, Japan, Jordan, Kenya, Liberia, Mauritania, Mongolia, Morocco, Netherlands, Norway, Senegal, Sweden, Tunisia, Turkey, United Kingdom, United States of America, and Zambia. Together, these nations represent around 34% of the global population and 51% of global greenhouse gas emissions, and contribute to around

Embodied Carbon. Ming Hu, Oxford University Press. © Ming Hu (2025). DOI: 10.1093/9780198944485.003.0003

64% of the global domestic product [4]. However, despite these commendable commitments, the effectiveness of these strategies in mitigating climate change is yet to be seen.

Historically, carbon reduction efforts in the building sector have concentrated on the operational phase, primarily due to its higher proportion of life-cycle emissions. This focus has overshadowed the significance of embodied carbon—the emissions associated with materials and construction processes—which is no longer negligible [5]. As buildings become more energy-efficient and operational carbon decreases, the relative importance of embodied carbon is rising dramatically. Recent studies project that embodied carbon, which currently constitutes about 20–50% of a building's life-cycle emissions, could escalate to account for as much as over half or, in some net-zero energy buildings, virtually all of the emissions [6]. These findings highlight a critical shift: As we progress towards energy efficiency, the battleground for carbon reduction is moving from the operational to the embodied, a domain still fraught with regulatory and research gaps.

The current narrative surrounding embodied carbon is one of underestimation and insufficient regulatory attention. Nevertheless, as the scales tip towards a preponderance of embodied carbon in highly energy-efficient buildings, its importance becomes increasingly paramount. The discrepancy in mitigation strategies for embodied carbon across various regions reflects not just a gap in practice but an opportunity for innovation and concerted global effort.

3.1.2 The significance of embodied carbon in the real world

The impacts of embodied carbon are paramount in the discourse on environmental sustainability, particularly within three distinct facets of the building sector. *First*, the embodied carbon is not a small amount. In the United States, for instance, the building sector was responsible for 36% of the nation's carbon emissions in 2022, with embodied carbon alone accounting for 11% of these emissions, according to the United States Green Building Council (USGBC) [7, 8]. Similarly, in the United Kingdom, the production of construction materials alone contributed 5–6% of the country's total emissions [9], and the total embodied carbon in the building sector is 20% [10]. Table 3.1 details the percentage of building and construction sectors' embodied carbon contributions to total national emissions. In most developed countries, embodied carbon typically accounts for 11 to 20% of total emissions. In contrast, developing countries exhibit higher embodied carbon percentages, a trend driven largely

Table 3.1 Embodied carbon emission percentage in existing building and construction sector

Country	Embodied carbon percentage (%)
United States	11 [8]
Netherland	11 [11]
Switzerland	20 [12]
France	25 [13]
Finland	15 [14]
United Kingdom	20 [10]
China	17 [15]
Brazil	30 [16]
India	40 [17]
Kenya	10 [18]

by ongoing construction activities aimed at expanding their building and infrastructure stock.

Second, as international regulations and policies increasingly emphasize operational energy efficiency in buildings, the significance of embodied carbon is becoming more pronounced [19]. In scenarios where buildings achieve net-zero operational energy—emitting no carbon during use—the focus inevitably shifts to embodied carbon, which could then account for the totality of a building's emissions throughout its life cycle.

Last, while enhancing energy efficiency in building operations is essential, it is crucial to address the carbon emissions associated with the materials used in renovations. Frequent refurbishments, often driven by the use of low-quality materials and construction methods, can significantly increase a building's overall carbon emissions. Therefore, reducing the embodied emissions in renovation materials and methods is imperative. Building designers play a pivotal role in this process, as their choices in materials and construction techniques directly influence the carbon footprint of buildings. To meet our broader energy conservation and sustainability objectives, reducing embodied energy in buildings is not merely desirable—it is essential.

3.2 Escalating emissions in the building sector and challenges

In 2022, CO_2 emissions from construction and building operations reached an unprecedented high, accounting for 37% of global emissions, roughly 10 gigatonnes (Gt) [2]. This rise is primarily attributed to emissions from building operations and the manufacture of key materials such as cement, steel, and aluminium, which alone contributed 2.5 $GtCO_2$ [2]. Additionally, the production of bricks and glass significantly impacted emissions, adding approximately 1.2 $GtCO_2$ [2]. Notably, these figures represent only the embodied carbon from raw material extraction to production (modules A1 through A3).

This assessment, however, captures just a segment of the materials involved. Including emissions from a broader array of building materials such as plastics, foams, fabrics, and various construction components reveals an even grimmer picture of the CO_2 impact. This revelation underscores the urgent need for substantial changes in construction practices and material usage, highlighting a critical need for education and professional focus within the built environment field. The challenge is twofold: managing rising emissions while meeting the intensifying requirements of climate change mitigation. The increase in emissions from the building sector is fuelled by urban expansion and the limited use of low-carbon materials. Let us delve deeper into these elements.

3.2.1 Urban expansion

Since 2010, the global footprint of built spaces has spiked by 31%, now surpassing 250 billion square meters, with residential constructions making up the majority, around 80% [20]. Curiously, while energy efficiency improvements have led to a decrease in energy use intensity since 2015, emissions from building operations have climbed by 5.4% [20]. The growth of built areas is a primary factor driving up energy demand for both building operations and the production of construction materials.

Between 2021 and 2022, global building floor space grew by 2%, with energy demand rising by 1%. The rate of construction varied globally in the years that followed. The United States saw consistent growth in construction activities among G20 countries, while countries in the Global South, including those in Africa, continued to see population and construction booms. The expected 15% increase in building floor area by 2030 will add roughly 40 billion square meters to the global built environment—equivalent to five times Indonesia's current built area [20]. The cost of construction is anticipated to jump from $9.7 trillion in 2022 to $13.9 trillion by 2037, marking a $4.2 trillion increase [21]. This surge is more than a set of statistics; it symbolizes the dynamic evolution of our cities and infrastructures. The looming expansion accentuates the critical need for sustainable practices to be woven into the fabric of the construction industry.

3.2.2 Low-carbon material

The second key factor is access to low-carbon and bio-based materials, which presents a complex challenge that is deeply intertwined with the uneven distribution of natural resources. Take wood, for instance, which has become a focal point in the drive toward bio-based building materials. Developed countries in the Global North have placed significant emphasis on the adoption of wood, particularly mass timber, as a strategy for diminishing embodied carbon emissions within the building sector [22]. Proponents have even suggested that engineered timber could transform the global building stock into a carbon sink to counteract the climate crisis [9].

However, this approach is flawed for a couple of reasons. Firstly, there's a glaring mismatch between supply and demand on a global scale. According to research conducted by Kraxner et al., the majority of sustainable forest areas suitable for harvesting wood for construction are located in the Global North, specifically in North America and Europe [23]. This is despite the fact that these regions have a relatively low rate of new constructions [24]. Meanwhile, the burgeoning building stock is set to double, primarily in developing countries of Africa and Asia—regions poised for significant population growth [24]. The demand for wood as a primary building material outstrips supply in these regions with the sharpest population increases.

Consequently, there is an imperative to move beyond the one-size-fits-all mindset. Alternative low-carbon options should be explored—solutions that leverage locally sourced materials and traditional construction knowledge, which are cost-effective, practical, and adaptable to various conditions. The indigenous building techniques, using materials available in the local environment, offer a wellspring of knowledge and a potential roadmap for sustainable construction practices tailored to specific regional needs. In Chapter 8, we will focus on low-embodied carbon materials that have been used in traditional building materials in the Global South.

3.3 Global supply chain vulnerability

Recent geopolitical tensions have significantly disrupted the economic landscape, leading to a cascade of challenges with profound implications for the construction sector, especially in the Global South (emerging and developing economies). Economic uncertainty has become a prevailing climate, while constraints tied to this instability have led to soaring commodity prices, further exacerbating the situation. The result is a domino effect where disrupted supply chains hinder the availability of construction materials, inflating costs across the board. In emerging and developing economies, there has been a marked slowdown in building construction economic activities due to the pressures of increased borrowing costs and builders facing higher costs of construction in labour and materials.

These supply chain disruptions can occur at various stages, including inbound material flow from suppliers, internal production processes, and outbound material flow to customers [25]. The construction industry is particularly vulnerable to disruptions, which can result in significant time and cost over-runs [26]. Examples of disruptions in construction supply chains include increased levels of non-conformance, delivery delays, and other supply disruptions, leading to inflated costs and project delays [27]. For example, during the pandemic, the lead times for bar joists and steel decks extended to 10–14 months [27].

This turbulent economic environment has particularly impacted the construction sector in developing regions of the Global South. The tightening of global financial conditions has manifested in increased borrowing costs. Builders and developers are now grappling with the dual challenge of securing financing at reasonable rates and contending with the inflated expenses for both labour and materials. In some cases, the cost of construction has risen to the point where projects are no longer financially viable, stalling development and contributing to a slowdown in construction economic activities.

The implications of these economic stressors are multifaceted. On one hand, they threaten the progress and sustainability of construction projects, which are vital for economic growth and development. On the other hand, they underscore the vulnerability of global supply chains and the need for resilience in construction practices. The urgency for local sourcing of materials and labour, along with the adoption of sustainable and cost-effective construction methods, has never been more apparent.

3.4 International regulation of embodied carbon

Initiating a new chapter in environmental regulation, the European Commission, as of 2022, has formally recognized embodied carbon within its legislative landscape. The sustainable finance taxonomy, effective from January 1, 2022, guides investments towards business practices beneficial to climate action [28]. This directive mandates that new construction projects over 5,000 square meters undergo an extensive review of their life-cycle global warming potential (GWP), measured by kg CO_2eq emission, ensuring transparency for stakeholders [28].

An upcoming revision to the Energy Performance of Buildings Directive (EPBD) will, from 2027, require all new buildings exceeding 2,000 square meters to complete a life-cycle assessment (LCA), fully documenting the carbon impact from inception to potential deconstruction. These findings will then be integrated into the building's energy performance certificate, with plans to broaden this mandate to smaller buildings by 2030, emphasizing carbon sequestration potential. Additionally, the EPBD calls for member states to develop tailored strategies to curtail embodied carbon in renovations and to optimize carbon uptake. This prospective policy is awaiting deliberation in the European Parliament and may evolve before adoption.

Following those published regulations and policies, the movement to control and diminish embodied carbon has gained momentum across Europe, with France pioneering stringent national policies. By the close of 2023, a collective of European nations, including Denmark, Finland, France, the Netherlands, and Sweden, had advanced regulations that encompass the complete life-cycle carbon emissions, showcasing a concerted effort to address the multifaceted nature of building-related carbon output. The subsequent section aims to elucidate the global landscape of embodied carbon regulation, with a focus on the comparative analysis of policy development across four representative countries from Northern Europe, Western Europe, North America, and Asia. The four countries are Finland, France, the United States, and China. This approach will facilitate a comprehensive understanding of the regional variations and the respective responses to the challenges posed by embodied carbon in the construction and manufacturing sectors.

3.4.1 Northern Europe

In August 2019, the Nordic Prime Ministers from Denmark, Finland, Iceland, Norway, and Sweden unified under a bold vision: to position the Nordic region at the forefront of global climate action and sustainable societal development [29]. This resolution was particularly focused on the construction sector, highlighted by the Nordic Carbon Neutrality Declaration on low-carbon construction, endorsed in October 2019 [30]. This declaration represents a concerted commitment to a standardized life-cycle approach to building sustainability, spearheaded by Finnish and Swedish authorities at the Nordic Climate Forum for Construction.

A pivotal survey distributed among approximately one-hundred forum attendees sought perspectives on standardizing LCA practices for buildings. The response was overwhelmingly positive, with 97% supporting unified LCA definitions within regulatory structures and 93% advocating for methodological uniformity to establish common LCA process boundaries [31]. Additionally, 90% backed the creation of a Nordic database to aggregate environmental data on construction materials [31].

Despite this consensus on the need for consistent LCA methodologies, actual practices in assessing embodied carbon remain diverse. Finland's new Building Act, enacted on 1 March 2023, exemplifies comprehensive climate action integration within building legislation. Among the Nordic countries, as demonstrated in Table 3.2, Finland considers a broad scope of life-cycle stages in its system boundaries. Conversely, Sweden's requirements encompass only stages A1 to A5, while Demark only requires A1 through A3 assessment, and Norway is in between, requiring A1 through A3 and B4, B5 assessment. In terms of included building system, Denmark and Finland, unlike Norway and Sweden, include building service systems, such as heating, within their physical system boundaries [29].

In addition, Finland has taken significant strides in creating a clear framework for reducing embodied carbon, as seen in its comprehensive 'Method for the Whole Life Carbon Assessment of Buildings' [32]. It is one of the most detailed and easy-to-use roadmaps for reducing embodied carbon produced at the national level [32].

Other Nordic countries have also established specific requirements. For example, the new Danish requirements (starting from January 2023) require that all new buildings document their environmental impact over a lifespan of fifty years through LCA calculation and that new buildings above 1000 square meters comply with the limit value of 12 kg CO_2 equivalent per square meter per year [33]. Sweden has introduced the Act on Climate Declarations for New Buildings, which requires developers to calculate and file the embodied carbon emissions for the upfront stages of a building's life cycle, including material production and construction (A1–A5).

3.4.2 France

The building sector in France plays a substantial role in the country's environmental footprint, responsible for roughly 25% of France's total green-

Table 3.2 Life-cycle stage boundaries in Nordic assessment schemes

Country	Production			Construction		Use							End of life				
	A1	A2	A3	A4	A5	B1	B2	B3	B4	B5	B6	B7	C1	C2	C3	C4	D
Demark	X	X	X														
Finland	X	X	X	X	X			X	X		X		X	X	X	X	X
Iceland	X	X	X	X	X										X	X	
Norway	X	X	X						X	X							
Sweden	X	X	X	X	X												

Notes: X = Included.

house gas (GHG) emissions and 44% of its energy usage. In the case of new energy-efficient buildings with low operational energy use, 60 to 90% of emissions comes from the construction and demolition phases [13]. In addressing this, France has ventured into new regulatory territories for carbon mitigation. The RE2020 regulation, effective from 2022, sets forth progressive caps on GHG emissions that will tighten over time, aiming for a systematic transition to low-carbon construction materials. This regulation encompasses both operational and embodied carbon, with an eventual goal of cutting emissions by 30% compared to 2022, translating to limits of 415 kg CO_2/m^2 for single-family homes and 490 kg CO_2/m^2 for apartment buildings by 2031 [13]. According to the fifty-year life span, this standard can be translated to 8.3 and 9.8 kg CO_2 equivalent per square meter per year, respectively, which is slightly lower than Denmark's goal.

In pursuit of these targets, France has enacted specific mandates governing material selection in construction. A key legislative piece demands that from 2022, all state-funded new public buildings must incorporate at least 50% timber or other natural materials, such as hemp or straw [34]. Supporting this shift, the French government is investing €420 million in the development of bio-based materials and the associated industrial technologies [35]. France is recognized as a leading nation in the domain of embodied carbon management, boasting the most extensive collection of Environmental Product Declarations (EPDs), as evidenced by the centralization of data within the INIES database. This database is regulated by the Alliance HQE-GBC, an authority that also oversees the Haute Qualité Environnementale (HQE) building certification [36]. The French framework includes two distinct EPD programmes: the Fiches de Déclaration Environnementale et Sanitaire (FDES), which pertains to construction materials, and the Profiles Environmental Product (PEP) EcoPassport, which is applicable to electrical devices, encompassing products like air conditioners and photovoltaic panels [36].

An analysis conducted by Dr Jane Anderson on the INIES database reveals that a significant majority of EPDs, amounting to 60%, are attributed to specific manufacturers. Of these, a substantial proportion originates from French manufacturers themselves, accounting for 60%, while other European entities contribute 25% [36]. In contrast, collective EPDs constitute a smaller portion, with 11% emanating from French trade associations and a mere 4% from broader European collaborations. These figures underscore a strong, geographically concentrated commitment to environmental certification within the French building industry [36].

3.4.3 United States

The United States is actively tackling embodied carbon in the building and construction sector through a suite of policies and initiatives designed to mitigate the environmental impact of construction materials. A cornerstone of federal policy in this area is the Biden–Harris Administration's Federal Buy Clean Initiative, introduced in 2021 as part of the broader Federal Sustainability Plan and Executive Order 14507 [37]. This initiative capitalizes on the significant purchasing power of the U.S. government to promote the use of construction materials with lower embodied greenhouse gas emissions. The interagency task force behind this initiative comprises members from key departments and agencies, including Commerce, Defense, Energy, Homeland Security, Housing and Urban Development, Health and Human Services, Interior, State, and Transportation, as well as the Environmental Protection Agency (EPA), General Services Administration (GSA), and the White House Domestic Climate Policy Office [37]. This collective effort underscores the federal government's dedication to guiding the industry towards sustainability. A particular focus has been placed on EPDs as a tool for enhancing transparency regarding suppliers' embodied emissions. To support this focus, the initiative is channelling funding, incentives, and technical assistance to boost the development and uptake of EPDs.

The Buy Clean Initiative gained further impetus with the Inflation Reduction Act (IRA) of 2022, a pivotal piece of legislation in regulating embodied carbon emissions [38]. This Act reinforces the commitment to environmental stewardship within the construction sector, promoting standardized, comprehensive methodologies for assessing, reporting, and validating the embodied carbon in construction materials and products. The IRA channels considerable resources to federal agencies, particularly the EPA and GSA, for the creation and management of programmes that identify, validate, and incentivize the use of lower-carbon construction materials.

In line with the IRA, the EPA is developing customized programmes to support a broad array of stakeholders, including state and local governments, tribal nations, industry manufacturers, procurement professionals, developers, and building contractors. These programmes are designed for the detailed measurement and documentation of embodied carbon, targeting a reduction in greenhouse gas emissions from construction materials through the implementation of EPDs and the establishment of a certification system for low-carbon materials [39].

In May 2023, the GSA launched the Low Embodied Carbon Pilot Program, a critical step towards implementing the IRA's provisions for the use of sustainable construction materials in GSA-funded projects [40]. This programme aims to stimulate the use of construction materials with a significantly reduced embodied carbon profile, leveraging cooperative ventures with the construction industry and enhancing government procurement processes. The Pilot includes a portfolio of eleven new construction projects, border infrastructure, modernization initiatives, and paving projects valued at over $300 million. These projects, varying in scope and phase, are spread across eight states and the District of Columbia. The Pilot specifically evaluates the GSA's interim IRA Low Embodied Carbon material requirements across different project types, with a focus on four key materials—asphalt, concrete, glass, and steel—identified by the EPA for their substantial potential to reduce embodied carbon.

However, despite these significant regulatory and policy-level efforts at the federal level, the United States still lacks a nationwide embodied carbon limit requirement, unlike Denmark and France. To explore how the U.S. might measure up to these nations, the author conducted a life-cycle assessment of new construction projects built according to the International Energy Conservation Code (IECC) 2021 standards. The findings indicate that the life-cycle embodied carbon for such buildings would range between 4.5 and 6.9 kg CO_2 equivalent per square meter per year, encompassing both the primary structure and building envelope system. If an additional 30% embodied carbon attributable to other building service systems, such as mechanical systems, is added, then the total life-cycle carbon is between 5.9 and 8.9 kg CO_2 equivalent per square meter per year—an outcome that approaches the targets set by Denmark and France, as illustrated in Figure 3.1.

In conjunction with federal initiatives, several state and local governments in the U.S. have implemented their own policies to address embodied carbon. Notably, California led the way with the Buy Clean California Act of 2017, mandating that from 2021 public projects exceeding $1 million utilize steel, flat glass, and mineral wool insulation that meet established embodied carbon limits, verified by EPDs [41]. The Act became fully operational in July 2022, introducing even more stringent requirements. Following California's lead, other states, such as Oregon, passed similar legislation with House Bill 4139 in March 2022, and Colorado enacted HB21-13-3 in 2021 [42], while a comparable bill in Washington State (ESHB1282) was unsuccessful in 2023 [43].

Some states have focused on regulating specific construction materials rather than a broader range of products. For example, New York State's Low Embodied Carbon Concrete Leadership Act (LECCLA), passed in 2021,

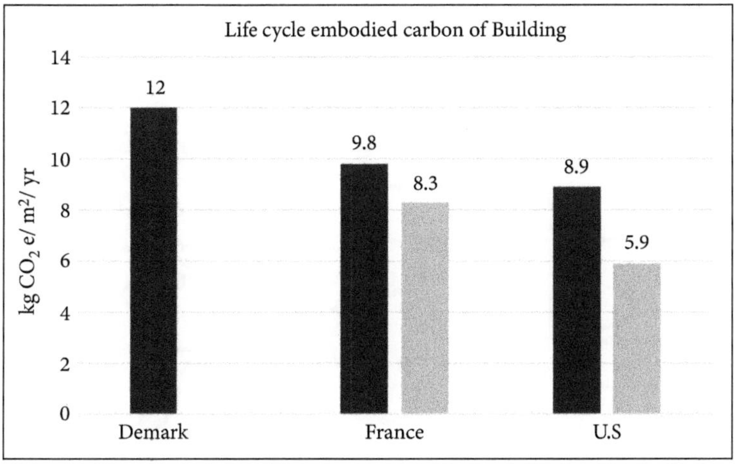

Figure 3.1 Life-cycle embodied carbon benchmark.

targets the public procurement of low-carbon concrete, with New Jersey enacting its version of the legislation in 2023 [44]. At the municipal level, Marin County in California was the pioneer in 2019, becoming the first U.S. county to embed requirements for low embodied carbon concrete in their building code, applicable to all construction projects, including privately funded ones [36]. This could be achieved by providing an EPD demonstrating a low carbon coefficient or by reducing the cement content in concrete, verified through batch receipts. Denver and Ann Arbor have also introduced amendments and resolutions to promote the use of materials with reduced embodied carbon. The Carbon Leadership Forum notes that over sixty cities and counties across the U.S. have adopted some form of regulation or policy aimed at curtailing embodied carbon [45].

Figure 3.2 illustrates the layered regulatory actions unfolding across federal, state, and local jurisdictions in the United States, reflecting a dedication to mitigating embodied carbon in the building and construction sector. These measures champion the adoption of materials with reduced embodied carbon and propel a shift within the industry toward increasingly sustainable operations in the United States.

3.4.4 China

A 2023 analysis of embodied carbon emissions across various sectors in China conducted by Wang and colleagues has underscored the construction sector's prominent contribution to the nation's carbon footprint [46]. The research illuminated the substantial demand for construction, which notably drives

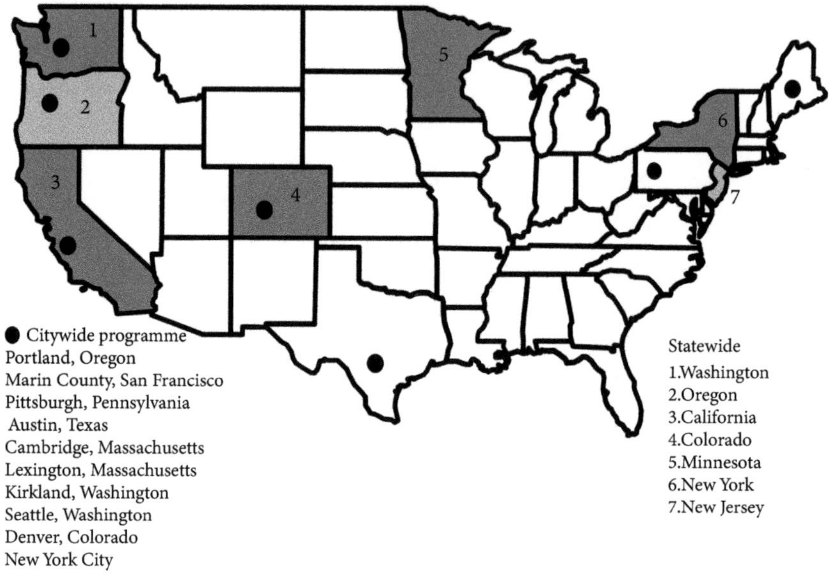

Figure 3.2 The U.S. embodied carbon programmes.

up emissions. In a progressive stride towards sustainability, China set forth a national goal in 2020 to peak carbon emissions by 2030 and achieve carbon neutrality by 2060. This pledge was strengthened by introducing the National Climate Change Adaptation Strategy 2035, which aims to establish an all-encompassing climate impact and risk assessment system by 2035, enhancing the country's pre-emptive measures and capacity to evaluate and adjust to environmental changes [47].

Despite these forward-looking measures, China's march towards urbanization marches on, with the construction sector as a pivotal growth engine. The sector's activities, fuelling the annual construction of an estimated 5.2 billion square meters of new buildings, result in the consumption of vast quantities of resources—specifically, 2 billion tons of cement and 600 million tons of steel [48]. Such intensive development contributed to the sector accounting for 45.8% of the nation's total energy consumption and 50.6% of total carbon emissions in 2019 [49]. These stark statistics serve as a clarion call for the adoption of more rigorous environmental standards governing construction practices and materials throughout China.

In line with the country's eco-friendly goals, 90% of China's commercial building owners have committed to adding at least one net- or near-zero energy structure to their portfolios in the coming ten years. This resolve is mirrored in China's significant slice of the global green-building pie—a 17.2% market share [50]. The government has joined forces with financial institutions to drive this change through green financial instruments such as green bonds, green credit, and green funds, aimed at nurturing the growth of green construction projects. By the end of 2021, China had seen its green loan balance skyrocket to RMB 15.9 trillion (US$2.23 trillion), a 33% increase from the year prior, cementing its position as a global leader in green finance [51]. These investments are strategically channelled into projects that focus on carbon emission reduction.

The Ministry of Housing and Urban–Rural Development (MHURD) has curated a list of fifty-one specific green building materials for certification under the Green Building Materials Product Certification Implementation Plan [52]. These materials are categorized into five key groups for better coherence and governance:

1. Building envelope structures and concrete
2. Doors, windows, curtain walls, and decorative elements
3. Waterproofing, sealing materials, and architectural finishes
4. Water supply, drainage, and sanitation systems
5. Heating, ventilation, and air conditioning (HVAC) systems, alongside solar energy solutions.

China's efforts to address embodied carbon challenges reflect a strategic commitment to ensuring that every aspect of building construction contributes to a sustainable future. This commitment is evident in China's robust policy framework. On 1 April 2022, MHURD implemented the General Code for Building Energy Conservation and Renewable Energy Utilization, mandating reductions in carbon emissions from the building sector [53]. This comprehensive regulation encompasses a wide range of activities, from retrofitting existing buildings to establishing new construction protocols, and from commissioning processes to building operations. The General Code sets ambitious energy conservation targets, stipulating that by 2025, all new urban buildings must meet green building standards and increase the use of renewable energy in buildings from 6% in 2020 to 8% by 2025.

Following the General Code, the Ministry of Finance and other agencies issued the Circular on 24 October 2022, expanding green building material procurement policies to forty-eight urban districts, including six pioneering cities, starting in November 2022 [54]. This policy applies to a broad range of public projects and mandates strict adherence to the Standards on

Governments' Procurement Demands Concerning Green Building and Green Building Materials across these cities.

Additionally, the Ministry of Industry and Information Technology (MIIT) released the Action Plan on Peaking Carbon Dioxide Emissions in the Building Materials Industry on 7 November 2022 [53]. This plan outlines fifteen key actions across five strategic areas to optimize resource management, improve solid waste utilization, and limit the use of high-carbon materials. It promotes advancing low-carbon technology, enhancing energy efficiency, and implementing carbon reduction measures. The plan also supports establishing a framework for encouraging the production of sustainable building materials, advancing China's sustainable development goals within the building materials sector.

It is worth noting that the current adoption of EPDs and other ecolabels in China is relatively low. However, under the influence of legislative pressure and financial support, a rapid increase in the implementation of EPDs and other programmes for reducing embodied carbon in the building and construction sectors can be anticipated.

3.5 Conclusion

Chapter 3 describes the escalating environmental burden posed by embodied carbon in the construction and building sector, emphasizing the critical need for immediate and sustained action. The built environment, while essential for human habitation and economic development, significantly contributes to global greenhouse gas emissions. With operational carbon becoming more efficiently managed, the focus must now shift to the often-overlooked embodied carbon embedded in building materials and construction processes. Global initiatives, such as France's legislative mandates for incorporating bio-based materials and the European Union's stringent regulations on life-cycle assessments, underscore the urgent need for comprehensive strategies to mitigate embodied carbon. Similarly, the United States' Federal Buy Clean Initiative and China's ambitious green finance mechanisms illustrate the growing recognition of embodied carbon's importance on the international stage.

Chapter 3 also highlights the alarming data which show that in highly energy-efficient buildings, embodied carbon could constitute the majority of a building's total emissions. This necessitates a paradigm shift in building design and construction practices, where architects and engineers must prioritize low-carbon materials and sustainable construction methods. Urban expansion and the increased demand for construction materials underscore the pressing need for innovative solutions. The use of locally sourced materials and traditional construction techniques can offer viable, sustainable alternatives, particularly in developing regions where the demand for new buildings is most acute. The challenge is twofold: reducing the embodied carbon in new constructions and addressing the carbon footprint of existing buildings through sustainable renovations.

In conclusion, Chapter 3 wraps up Part I by exploring past developments and emphasizing that achieving a sustainable built environment requires addressing embodied carbon with the same rigour as operational carbon. Global cooperation, leveraging of technological advancements, and enactment of robust policies are crucial to significantly mitigate the environmental impact of the construction sector. Significant emissions reductions depend on the widespread adoption of low-carbon materials and the implementation of stringent regulatory frameworks that promote sustainable practices. The time to act is now, as today's decisions will shape the carbon footprint of our cities for generations. Through concerted efforts, the construction industry can transition towards a more sustainable future, balancing development needs with environmental stewardship.

As we move into Part II, we will delve into the complexity and hidden inequities associated with embodied carbon. Chapter 4 will begin this exploration by examining the shifting resource and environmental burdens linked to global construction practices. It will focus on how raw material extraction in the Global South supports building activities in the Global North, and the resulting implications for both regions.

References

1. Pomponi F, Moncaster A. Embodied carbon mitigation and reduction in the built environment—What does the evidence say? Journal of Environmental Management. 2016 Oct; 181:687–700.
2. UN Environment Programme. 2023 Global status report for buildings and construction [Internet]. 2024 Mar [accessed 9 Apr 2024]. Available from https://www.unep.org/resources/report/global-status-report-buildings-and-construction
3. UNEP. 2020 Global status report for buildings and construction: towards a zero-emission, efficient and resilient buildings and construction sector [Internet]. United Nations Environment Programme; 2020. Available from https://globalabc.org/sites/default/files/inline-files/2020%20Buildings%20GSR_FULL%20REPORT.pdf

4. UN Environment Programme. The buildings breakthrough: global push for near-zero emission and resilient buildings by 2030 unveiled at COP28 [Internet]. 2023 [Accessed 18 Jun 2024]. Available from https://www.unep.org/news-and-stories/press-release/buildings-breakthrough-global-push-near-zero-emission-and-resilient
5. Hu M, Esram NW. The status of embodied carbon in building practice and research in the United States: a systematic investigation. Sustainability. 2021 Nov 23; 13(23): 12961.
6. Röck M, Saade MRM, Balouktsi M, Rasmussen FN, Birgisdottir H, Frischknecht R, et al. Embodied GHG emissions of buildings—the hidden challenge for effective climate change mitigation. Applied Energy. 2020 Jan; 258: 114107.
7. Verdinez D. U.S. Green Building Council and RMI partner to develop embodied carbon guidelines [Internet]. 2023 [Accessed 15 Apr 2024]. Available from https://www.usgbc.org/articles/us-green-building-council-and-rmi-partner-develop-embodied-carbon-guidelines
8. Huynh T, Magwood C, Olgyay V, Kerr L, Sullens W. Driving action on embodied carbon in buildings, RMI and U.S. Green Building Council (USGBC) [Internet]. 2023 Sep [accessed 15 Apr 2024]. Available from https://www.usgbc.org/sites/default/files/2023-09/driving_action_on_embodied_carbon_in_buildings_report.pdf
9. Rawlinson S, Weight D. Sustainability: embodied carbon. Building Magazine. 2007; 12: 88–91.
10. UKGBC. Embodied carbon modelling and reporting [Internet]. [Accessed 14 Apr 2024]. Available from https://ukgbc.org/resources/embodied-carbon-modelling-and-reporting/#:~:text=Embodied%20carbon%20also%20makes%20up,made%20across%20an%20asset's%20lifetime.
11. Dutch Green Building Council. Whole life carbon-position paper. Dutch Green Building Council; 2021 Sep.
12. Federal Office for the Environment. Interactive CO2 calculator on the federal government's geoportal [Internet]. [Accessed 15 Apr 2024]. Available from https://www.bafu.admin.ch/bafu/en/home/topics/climate/info-specialists/reduction-measures/buildings/calculator-co2-buildings.html
13. Agora Energiewende. Reducing embodied carbon in construction materials: RE2020 in France [Internet]. [Accessed 27 Mar 2024]. Available from https://www.agora-energiewende.org/international/success-stories/reducing-embodied-carbon-in-construction-materials-re2020-in-france-1#downloads
14. Climate roadmaps 20235. Low-carbon roadmap for the Finnish construction industry [Internet]. [Accessed 15 Apr 2024]. Available from https://www.climate2035.fi/roadmaps/construction-industry/
15. Lu H, You K, Feng W, Zhou N, Fridley D, Price L, et al. Reducing China's building material embodied emissions: Opportunities and challenges to achieve carbon neutrality in building materials. iScience. 2024 Mar; 27(3): 109028.
16. Lassio JG, Branco DC, Magrini A. Assessing the greenhouse gas emissions of buildings in brazil: a case study of a housing complex. Environmental Engineering and Management Journal. 2021; 20(8): 1225–36.

17. Pal Singh Raina G. Decarbonising the built environment in India [Internet]. London: Royal Institute of Chartered Surveyors; 2023 Nov [accessed 15 Apr 2024]. Available from file:///Users/minghu/Downloads/rics-cop28-policy-paper-india.pdf
18. Onyalo N, Tembo M. An evaluation of Kenya's climate change policy challenges for the building industry toward the realization of netzero emissions. International Journal of Computing and Engineering. 2023; 4(1): 42–70.
19. Hu M. Embodied carbon emissions of the residential building stock in the United States and the effectiveness of mitigation strategies. Climate. 2022 Sep 20; 10(10): 135.
20. IEA. Tracking clean energy progress 2023 [Internet]. 2023 [Accessed 10 Apr 2024]. Available from https://www.iea.org/reports/tracking-clean-energy-progress-2023
21. Economic Consulting Team. Global construction futures [Internet]. [Accessed 12 Apr 2024]. Available from https://www.oxfordeconomics.com/resource/global-construction-futures/
22. Pomponi F, Hart J, Arehart JH, D'Amico B. Buildings as a global carbon sink? A reality check on feasibility limits. One Earth. 2020 Aug; 3(2): 157–61.
23. Kraxner F, Schepaschenko D, Fuss S, Lunnan A, Kindermann G, Aoki K, et al. Mapping certified forests for sustainable management—a global tool for information improvement through participatory and collaborative mapping. Forest Policy and Economics. 2017 Oct; 83: 10–18.
24. Hamiltton I, Rapf O, Kockat D, Zuhaib D. 2021 Global status report for buildings and construction [Internet]. United Nations Environmental Programme; 2021 Oct. Available from https://www.unep.org/resources/report/2021-global-status-report-buildings-and-construction
25. Habermann M, Blackhurst J, Metcalf AY. Keep your friends close? Supply chain design and disruption risk. Decision Sciences. 2015 Jun; 46(3): 491–526.
26. Zainal Abidin NA, Ingirige B. The dynamics of vulnerabilities and capabilities in improving resilience within Malaysian construction supply chain. CI. 2018 Nov 6; 18(4): 412–32.
27. Fauzi A, Nazaruddin, Wibowo RP. Measurement of supply chain management performance on toll road projects with design and build contract using SCOR and AHP methods: a case study on Tebing Tinggi-Parapat Toll Road Section III Project. In: Murhadi WR, Anandya D, Darmasetiawan NK, Dyah Trisnawati J, Mahadwartha PA, Tandelilin E, editors. Proceedings of the 19th International Symposium on Management (INSYMA 2022) [Internet]. Dordrecht: Atlantis Press International BV; 2023 [accessed 10 Apr 2024]. pp. 1152–61. (Advances in Economics, Business and Management Research; vol. 223). Available from https://www.atlantis-press.com/doi/10.2991/978-94-6463-008-4_142
28. European Commission. EU Taxonomy Navigator [Internet]. [Accessed 28 Mar 2024]. Available from https://ec.europa.eu/sustainable-finance-taxonomy/home?reference=7.1
29. Kuittinen M, Häkkinen T. Reduced carbon footprints of buildings: new Finnish standards and assessments. Buildings and Cities. 2020; 1(1): 182–97.

30. Declaration on Nordic Carbon Neutrality [Internet]. 2020 [Accessed 27 Mar 2024]. Available from https://www.norden.org/en/declaration/declaration-nordic-carbon-neutrality
31. Swedish Life Cycle Center. Summary of round table discussions during Nordic Climate Forum for Construction [Internet]. 2019 [Accessed 27 Mar 2024]. Available from https://www.lifecyclecenter.se/projects/nordic-working-group-for-harmonization-lca-climate-and-buildings/
32. Ministry of the Environment. Method for the whole life carbon assessment of buildings [Internet]. 2019 [Accessed 27 Mar 2024]. Available from https://julkaisut.valtioneuvosto.fi/bitstream/handle/10024/161796/YM_2019_23_Method_for_the_whole_life_carbon_assessment_of_buildings.pdf?sequence=1&isAllowed=y
33. Nordic Sustainable Construction. Buildings' life cycle assessments gain ground in the Nordics [Internet]. 2023 [Accessed 27 Mar 2024]. Available from https://nordicsustainableconstruction.com/news/2023/january/denmark-introduces-co2-limit-for-new-constructions#:~:text=The%20new%20Danish%20requirements%20entail,per%20square%20meters%20per%20year.
34. Fabris P. France to mandate all new public buildings be 50% timber or other natural materials [Internet]. 2022 [Accessed 31 Mar 2024]. Available from https://www.bdcnetwork.com/france-mandate-all-new-public-buildings-be-50-timber-or-other-natural-materials
35. World Bio Market Insights. France to invest €420 million in bio-based industries [Internet]. 2022 [Accessed 31 Mar 2024]. Available from https://worldbiomarketinsights.com/france-to-invest-e420-million-in-bio-based-industries/
36. Anderson J. Reducing embodied carbon in the built environment: the role of Environmental Product Declarations. 2023; PhD thesis, the Open University.
37. Office of the Federal Chief Sustainability Officer. Federal Buy Clean Initiative [Internet]. [Accessed 12 Feb 2024]. Available from https://www.sustainability.gov/buyclean/
38. The White House. Building a clean energy economy: a guidebook to the Inflation Reduction Act's investments in clean energy and climate action [Internet]. 2023 Jan [accessed 25 Mar 2024]. Available from https://www.whitehouse.gov/wp-content/uploads/2022/12/Inflation-Reduction-Act-Guidebook.pdf
39. U.S Environmental Protection Agency. Grant program: reducing embodied greenhouse gas emissions for construction materials and products [Internet]. [Accessed 8 Apr 2024]. Available from https://www.epa.gov/greenerproducts/grant-program-reducing-embodied-greenhouse-gas-emissions-construction-materials-and
40. U.S. General Services Administration. Low embodied carbon pilot fact sheet [Internet]. 2023 [Accessed 26 Mar 2024]. Available from https://www.gsa.gov/real-estate/gsa-properties/inflation-reduction-act/lec-program-details/lec-pilot-fact-sheet#:~:text=GSA's%20first%2Dto%2Dmarket%20low,levels%20of%20embodied%20greenhouse%20gas
41. Real Estate Services Division. Buy Clean California Act (BCCA) requirements [Internet]. [Accessed 27 Mar 2024]. Available from https://www.dgs.ca.gov/RESD/Resources/Page-Content/Real-Estate-Services-Division-Resources-List-

Folder/Buy-Clean-California-Act-BCCA-Requirements#:~:text=The%20Buy%20Clean%20California%20Act%20(BCCA)%20(Public%20Contract%20Code,used%20in%20public%20infrastructure%20projects.
42. LPDD. Oregon buy clean legislation [Internet]. [Accessed 27 Mar 1925]. Available from https://lpdd.org/resources/oregon-buy-clean-legislation/
43. BlueGreen Alliance. Washington BlueGreen Alliance | 2023 legislative session wrap-up [Internet]. [Accessed 27 Mar 2024]. Available from https://www.bluegreenalliance.org/work-state/washington/#:~:text=Labor%20and%20environmental%20partners%20joined,conditions%20of%20certain%20construction%20materials.
44. State of New Jersey. New Jersey Senate Bill 287 [Internet]. [Accessed 27 Mar 2024]. Available from https://legiscan.com/NJ/text/S287/id/2476603
45. Carbon Leadership Forum. Policy tracking map [Internet]. [Accessed 27 Mar 2024]. Available from https://carbonleadershipforum.org/clf-policy-toolkit/#map
46. Wang C, Zhao L, Papageorgiou GN, Qian Y, Xue J, Li D. Embodied carbon emissions generated by international trade of China's light industry sector based on global supply chains perspective. Energy Strategy Reviews. 2023 May; 47: 101095.
47. National Climate Change Adaptation Strategy 2035 [Internet]. 2022 [Accessed 1 Apr 2024]. Available from https://cset.georgetown.edu/publication/national-climate-change-adaptation-strategy-2035/
48. 刘芸曼. 绿色建材大势所趋 产业升级何去何从? [Internet]. Industry observation. [Accessed 1 Apr 2024]. Available from http://www.ce.cn/cysc/00/18/index.shtml
49. Wu Y. China's green building materials sector: policies and investment opportunities [Internet]. China Briefing. 2022 [Accessed 1 Apr 2024]. Available from https://www.china-briefing.com/news/chinas-green-building-materials-sector-policies-and-investment-opportunities/
50. Weber L. China commits to green construction [Internet]. 2021 [Accessed 1 Apr 2024]. Available from https://www.fdiintelligence.com/content/opinion/china-commits-to-green-construction-80293
51. 《中国绿色金融发展研究报告2021》发布 [Internet]. 2022 [Accessed 1 Apr 2024]. Available from http://www.ce.cn/xwzx/gnsz/gdxw/202204/29/t20220429_37544452.shtml
52. 三部门关于印发绿色建材产品认证实施方案的通知 [Internet]. 2019 [Accessed 1 Apr 2024]. Available from https://www.gov.cn/xinwen/2019-11/08/content_5450047.htm
53. 四部门关于印发建材行业碳达峰实施方案的通知 [Internet]. 2022 [Accessed 1 Apr 2024]. Available from https://www.miit.gov.cn/zwgk/zcwj/wjfb/tz/art/2022/art_8f6d55dd58d64283937d7fb87e21b666.html
54. Ministry of Finance of the People's Republic of China. Circular on extending the implementation scope of policies on governments' procurement of green building materials to improve building quality [Internet]. 2022. Available from http://gks.mof.gov.cn/guizhangzhidu/202210/t20221024_3847589.htm

PART II
UNRAVELLING THE COMPLEXITY AND THE HIDDEN INEQUITIES

Chapter 4
Embodied carbon

An environmental justice perspective

4.1 Shifting resource extraction and environmental burden

The complexities of environmental injustice, particularly concerning the overlooked aspects of embodied carbon and its contribution to broader societal inequities, indicate a significant gap in our collective understanding. It is imperative to recognize the intricate ties between environmental justice, the carbon embedded within global supply chains, and the socioeconomic disparities they exacerbate. Failure to acknowledge these links can lead to a cycle of distributive injustice and the unintended transfer of emissions from one region to another, known as emission leakage [1].

In the context of global construction practices, the extraction of raw materials in the Global South to produce building products for use in the Global North represents a significant shift of environmental burden. This phenomenon is intricately linked with global trade dynamics, economic inequality, and environmental policy. When raw materials are extracted in the Global South, this life-cycle phase (A1) is associated with considerable environmental degradation, including habitat destruction, water pollution, and greenhouse gas emissions. However, the countries where these materials are sourced often do not directly reap the benefits of their use. Let's look at a specific example of bauxite.

Bauxite mining, particularly in Guinea (23% world's reserve), one of the poorest countries [2], offers a stark illustration of the environmental toll of raw material extraction, as illustrated in Figure 4.1. This ore, essential for aluminium production [3], is plentiful in the region, but the cost to the environment is high [4]. The process begins with deforestation and topsoil removal, which are fundamental to mining operations. These actions strip away the topsoil and natural vegetation, laying waste to habitats and destabilizing ecosystems that have developed over millennia. The extraction doesn't stop at surface damage; it also leads to soil erosion, water contamination, and the release of pollutants into local water systems, endangering both terrestrial

Embodied Carbon. Ming Hu, Oxford University Press. © Ming Hu (2025). DOI: 10.1093/9780198944485.003.0004

Figure 4.1 (left) Former Bauxite mine.

and aquatic life. The carbon emissions from processing bauxite contribute to the broader climate crisis, adding an invisible yet profound layer of impact [5]. Why is bauxite mining booming? In addition to the steady demand for aluminium products, the increased push for electric vehicles (EVs) in the Global North is driving a new 'gold rush' for various metals, including bauxite. EVs require six times the mineral input by weight compared to conventional cars. As the Global North accelerates EV adoption, the costs and unintended consequences of securing these minerals are often overlooked [6].

This environmental onslaught of strip-mining transforms once-fertile land into barren terrain, heightening the risk of drought and the occurrence of soil cracks. The loss of the natural topsoil ecosystem, which cannot be easily restored after strip mining, leaves the earth dry and infertile. As dust rises from these parched lands, it exacerbates air pollution and further sterilizes the soil. For the local communities whose lives and livelihoods are intrinsically linked to the land and its boundary, the repercussions are dire. The natural resources they depend on are vanishing, undermining their traditional ways of living and compromising their future. While economically significant, bauxite mining casts a long shadow over the well-being of these communities and the integrity of their environment. Reports from the Guinean government and corroborations from rights activists and local inhabitants indicate that the compensation for these acquisitions, if provided at all, has been minimal. Looking ahead, projections from a governmental study suggest that bauxite mining will obliterate over 200,000 acres of farmland and 1.1 million acres of natural habitat within two decades—an expanse nearly equivalent to the size of Delaware [6]. The implications of such extensive land use conversion are profound, with irreversible losses to both agricultural productivity and biodiversity.

Despite the consequences, Guinea and other bauxite-rich countries such as Australia and Indonesia do not necessarily benefit from the end-use of these materials. Once extracted, bauxite is shipped to manufacturing countries to be turned into consumer products like aluminium window frames. According to 2022 data, China is the largest importer of bauxite, accounting for 75.2% of global bauxite imports, with Guinea exporting 64.3%, followed by Australia (12%), and Indonesia (9.33%) [7].

In China, raw bauxite is processed into aluminium window frames, an energy-intensive procedure that contributes significantly to the embodied carbon of the final product—87% in an insulated window [8]. This high carbon and energy footprint is primarily due to the substantial energy requirements involved in the initial smelting of bauxite. The Bayer process, which involves calcination of bauxite ore at high temperatures in large electrolytic cells, typically around 950°C (1,742°F), necessitates significant electrical energy input. Furthermore, after smelting, the aluminium often undergoes extrusion, where it is heated to around 500°C (932°F) and forced through a die to create the desired shapes and profiles for window frames, adding to energy consumption [9]. Subsequent surface treatments like anodizing or painting, which improve the appearance and durability of the frames, also involve additional energy-intensive heating, chemical treatments, or curing operations.

In addition to the carbon emissions from intensive energy use, the production process causes substantial environmental damage, such as air and water pollution and land degradation. The Bayer process, an industrial method for refining bauxite ore to produce alumina, which is the main precursor for aluminium production, releases particulate matter, nitrogen oxides, and sulphur dioxide into the air, contributing to respiratory issues and acid rain. Other liquid wastes containing dissolved alumina, caustic soda, and other compounds can further pollute water resources. Bayer process also requires large amounts of water and generates significant quantities of alkaline red mud waste as illustrated in Figure 4.2, which can contaminate water sources if not properly managed.

Land degradation and disposal issues arise from the large areas required for red mud waste disposal, a by-product of the alumina production process,

Figure 4.2 Pit with a puddle in red mud from iron ore.

Reproduced with permission from Dmytro via Adobe Stock, https://stock/adobe.com file#: 444558014.

leading to potential soil contamination. Improper storage and management of red mud can result in leaks, spills, and the release of toxic substances into the environment [10, 11]. Exposure to red mud can cause skin irritation, burns, and eye damage, and if improperly managed, it can contaminate water sources, posing broader health risks [12]. All these health risks and environmental damages are borne by workers and local communities in China. At the same time, the final product is shipped to Global North countries by demand, and profits are shared among stakeholders in developed countries. These products are used in the construction of new buildings or the renovation of old buildings. According to 2022 data, the largest importers of aluminium doors, windows, frames, and thresholds are the United States (20.8%), Germany (7.51%), France (7.13%), Japan (5.62%), the United Kingdom (5.01%), and Australia (4.85%). Together, these countries account for over 50% of the product demand [13].

The profits from manufacturing these high-value products predominantly remain in the Global North, where the building industry utilizes aluminium to create sophisticated structures. Meanwhile, the Global South countries, which bear the environmental and social impacts of aluminium extraction, receive a relatively small portion of the economic value generated from the final products [6]. This disparity underscores the unequal distribution of benefits and burdens in the global aluminium supply chain.

The transition from raw material extraction in the Global South to product consumption in the Global North has shifted the environmental burden from consumers to manufacturers. This shift is driven by asymmetrical global economic conditions, where the demand for raw materials in the Global North influences extraction activities in the Global South [14]. Studies have emphasized the significant environmental impact of this process, particularly in terms of biodiversity loss, heritage destruction, and extreme weather conditions [15].

Moreover, this geographical shift in burden is often exacerbated by the fact that environmental regulations may be stricter in the Global North, leading to the outsourcing of more polluting stages of production to countries with less stringent controls. This regulatory discrepancy can result in a form of environmental arbitrage where manufacturers and suppliers take advantage of the lower compliance costs in the Global South, effectively transferring the burden of pollution and environmental management away from the point of consumption to the point of extraction and initial processing.

4.2 Energy use intensity and emission paradox

The consequences of environmental burden shifting extend beyond ecological harm, leading to social and health issues within affected communities. While the Global North enjoys the advantages of construction and development, they often do not face the direct environmental and ethical consequences. Outwardly, these developed nations may seem committed to reducing environmental impacts, but this facade masks the reality that their progress is often at the expense of the Global South's increased energy consumption and environmental degradation.

Updated data from 2022 paints a stark picture: As listed in Table 4.1.

China's total energy use outstripped that of the United States, with figures standing at 159.39 and 95.91 exajoules (EJ), respectively [16]. However, when assessing energy use per capita, an individual in the U.S. consumes over twice as much as someone in China—283.50 compared to 111.80 GJ/person [16]. India, as the third-largest energy consumer, has a per capita energy consumption of only 25.70 GJ/person, a mere fraction of U.S. levels [16].

This pattern isn't unique to the United States. Developed countries in the Global North consistently show higher per capita energy use compared to their Global South counterparts. Energy consumption per capita is a significant measure since it correlates with living standards and the Human Development Index (HDI) [17]. High energy use often reflects access to modern amenities, healthcare advancements, and an overall improved quality of life [18, 19]. Table 4.1 in the discourse lays bare this disparity. The magnitude of the difference in per-person energy use between the highest Global North country (Canada) and the lowest Global South country (India) is fourteen times, underscoring the need for equitable living standards across the globe.

Moreover, we should be aware that per-person emissions listed here are production-based emissions that account for the emissions generated within a country's own borders. Since a large portion of the products are generated for export to meet the consumption demand in other countries, then consumer-based emissions can more accurately reflect the countries' emission contributions. Table 4.1 clearly shows that Global North developed countries have higher per capita emissions than Global South developing countries.

Table 4.1 Overall energy use and per-person energy use (2022)

	Country	**Per-person energy use 2022 (GJ/cap/yr) [16]**	**Per-person emission (ton/cap/yr) [20] 16/09/2024 20:29:00**	**Total energy use (EJ) [16]**
North	United States	283.50	14.44	95.91
	Japan	143.90	8.61	17.84
	Canada	367.80	15.22	14.14
	Germany	147.50		
	France	129.80	4.76	8.39
	United Kingdom	108.40	5.00	7.31
	Netherlands	201.60	7.79	3.54
	Sweden	215.70	3.69	2.28
	Norway	348.90	7.62	1.90
	Finland	212.30	6.64	1.18
	Switzerland	120.10	4.11	1.05
	Australian		15.12	5.98
South	China	111.80	8.85	159.39
	India	25.70	1.91	36.44
	Brazil	62.30	2.15	13.41
	Mexico	68.40	3.56	8.73
	Vietnam	46.70	3.27	4.59
	Thailand	70.60	4.06	5.06

Therefore, it's unjust to curtail the growth in energy use and associated emissions in developing countries without addressing this inequity. Only looking at the total energy use presents incomplete information and disguises the real problems. A comprehensive strategy for reducing carbon emissions cannot be achieved without considering the energy use and emissions on a per capita basis. For a genuine global reduction in carbon emissions, Global North developed countries must consider substantial cuts in their own per capita energy use and emissions, allowing for the increase of per capita in developing countries, in order to achieve the balance.

4.3 The unseen carbon cost: darker side of green

As explained in Chapter 3, the regulation of embodied carbon varies widely across the globe, with most practices being relatively new. These practices are driven by diverse regulatory, technical, and economic factors, which can lead to conflict and competition among countries. In this chapter, we will focus on the current imbalance in carbon emission measures and the reality of emission reduction. To begin, let us examine two paradoxical concepts that have emerged in the field of sustainability with the aim of reducing carbon emissions: renewable energy generation and preservation.

4.3.1 New is good? Technology is good?

In almost all carbon emission reduction or carbon neutrality plans, technological advancements have been treated as keys to solutions. Solar photovoltaic panel, wind turbine, and geothermal and recent carbon capture ignite people's high interest and hope. But are they really the saviours? Take, for example, the case of household solar photovoltaic (PV) panels, which have been lauded for their ability to produce electricity on-site, thereby reducing operational energy use and associated emissions. They also hold the potential for homeowners to return excess energy to the grid through mechanisms like net metering, further displacing the need for fossil fuel-based energy. However, the manufacturing landscape of solar panels reveals underlying concerns. Most solar panel components are not manufactured in Global North countries but are produced predominantly in China. In fact, according to 2021 figures from the International Energy Agency, China accounts for three-quarters of the world's solar panel production, including a higher percentage of the essential components such as cells (85%), wafers (97%), and polysilicon (79%). This disparity has illustrated the complex global interdependencies that can influence environmental sustainability and economic development [21].

Solar panel production is intertwined with environmental and health concerns due to its reliance on toxic materials like lead, cadmium, and sulphur hexafluoride [22, 23]. If not properly managed, these substances pose significant risks, particularly when panels are decommissioned and disposed of without proper safeguards [24]. The industry's supply chain is also challenged by the necessity for critical raw materials and the handling of toxic elements integral to solar panel construction. Moreover, there's a discrepancy in who bears the burden of these impacts. While consumers in the United States reap the benefits of clean energy and reduced emissions, the environmental and carbon costs accrued during raw material extraction and panel production are often outsourced to other regions. This geographic shift in the impact highlights an issue of equity and responsibility, where the advantages of green energy are enjoyed locally, but the ecological toll is borne elsewhere.

The solar panel industry is simultaneously scrutinized for its labour practices. Unstable employment and unexpected layoffs, particularly noted in Germany, spotlight the economic volatility linked to this sector [25]. This instability extends to places like China, where the workforce often contends with low wages and precarious work conditions [26]. Such issues are critically evident in the production of thin-film solar photovoltaic panels, where workers deal with dangerous substances like cadmium [27].

Research within the United States, such as studies by Cyrs et al. and Sinha et al., has frequently aimed to demonstrate the safety of solar panels, suggesting that the toxicity levels of materials used are negligible [28, 29]. However, those studies have been critiqued on two significant grounds. Firstly, the use of American data to simulate toxic exposure in China is questionable, casting doubts on the validity and applicability of such studies. Secondly, these studies have primarily focused on end-stage solar panels rather than the entire manufacturing and material extraction process where most of the toxic exposure actually occurs. Furthermore, there is increasing epidemiological evidence linking cadmium exposure, often associated with solar panel manufacturing, to a spectrum of cancers, challenging the notion that solar panels are harmless [30]. This situation underlines a deeper environmental injustice, compounded by flawed research methodologies that overlook the true breadth of the issue. In addition, while solar energy is celebrated for its green credentials, the industry's darker side is its hazardous waste generation. This poses immediate environmental risks and looms as a legacy concern for future generations. The pressing need for sustainable waste management solutions and labour reforms within the solar energy sector is paramount to ensure that the path to clean energy does not perpetuate other forms of environmental degradation [31, 32, 33].

Solar panels are just one example. As we delve into the paradoxes of solar energy and recognize the environmental challenges of modern green technologies, we see a reflection of broader issues within the sustainable development paradigm. The lessons learned from the complexities of embodied carbon in the energy sector provide a valuable framework for addressing similar challenges in urban planning. Just as the pursuit of renewable energy demands careful consideration of its full life-cycle impacts, so the stewardship of our built heritage requires a holistic approach. Within this context, we transition from discussing the delicate balance of ecological responsibility and industrial progress to exploring how these principles apply to the preservation of historical architecture and urban development. In section 4.3.2, we extend the conversation to the intricate ways in which our preservation efforts can contribute to—or conflict with—our environmental and social equity goals, underlining the necessity for strategies that encapsulate the integrity of our past while embracing the sustainability imperatives of our future.

4.3.2 Old is good? Preservation is good?

The discourse on heritage preservation is deeply entrenched in the societal narrative, stretching back not just decades but centuries. As time has progressed, the very concept of 'heritage' has evolved to include not only the storied stone structures of the eighteenth century but also the architectural marvels of the mid-twentieth century and onwards. The realm of preservation has similarly broadened to cover a spectrum of activities from adaptive reuse and renovation to simple maintenance—all of which are integral to upholding cultural continuity and the unique character of places within our urban tapestry. These practices are often woven into zoning and development frameworks, reflecting their importance in our collective approach to urban planning [34]. Furthermore, these preservation efforts play a crucial role in the environmental narrative, offering pathways to significantly reduce embodied carbon emissions through the careful stewardship and thoughtful retrofitting of existing buildings, as well as innovative approaches like vertical infill. Many practitioners and scholars have written about the benefits of preservation, conservation, and adaptive reuse as very effective ways to mitigate climate change through reducing unnecessary embodied carbon emissions, including the author's publication [35, 36, 37, 38].

Yet, preservation's legacy is not without its complexities. Traditionally seen as a vehicle for conservation, it has at times been wielded as a contentious instrument by those resisting change, particularly in affluent neighbourhoods, as a problematic tool of NIMBYism [39]. NIMBYism, an acronym for 'Not In My Back Yard', is a term that describes the phenomenon where residents of a community oppose new developments or projects that they perceive as potentially disruptive or detrimental to their local area. Although the individuals involved might support the project in principle—like affordable housing or renewable energy facilities—they object to its placement or implementation in their immediate vicinity due to concerns about decreased property values, increased traffic, or changes in community character [39]. Under the banner of heritage preservation, there have been instances where necessary initiatives—aimed at addressing critical housing shortages or creating affordable housing solutions—have faced opposition, revealing the delicate balance between preserving the past and nurturing a sustainable, inclusive future.

For example, in 2017, Seattle faced such a predicament when its historic preservation board vetoed a 200-unit apartment complex for exceeding the height of neighbouring historic buildings, despite compliance with local zoning laws. Consequently, the site remained a parking garage, squandering the opportunity for new housing next to a major transit hub and exacerbating the local housing crisis [40]. Similarly, Los Angeles witnessed a situation where the historical preservation argument became a veil for NIMBYism. Residents of an existing high-rise leveraged the protection of a clock on an older building to oppose the development of another skyscraper nearby. This scenario is highlighted by Urbanize LA, which stated, 'How do you fight the project next door while avoiding the dreaded NIMBY label? Just call it Historic Preservation' [41].

This ongoing struggle between preserving architectural heritage and accommodating growth is particularly pronounced in urban areas facing housing insufficiencies. When historic preservation unintentionally impedes housing development, it can exacerbate the shortage, inflating rental prices and displacing lower-income residents. The challenge lies in achieving a harmonious balance that respects history without arresting the development of the cityscape, especially for those low-income families.

In addition, investment in historic districts, while aimed at preserving cultural legacies, often inadvertently spurs gentrification, leading to the displacement of long-standing communities around the globe [42, 43, 44]. This unintended consequence of preservation efforts raises concerns about the equitable distribution of benefits and burdens. As a result, there's a growing shift towards equity-centred preservation practices, acknowledging the complex interplay between safeguarding history and fostering social justice [45, 39].

Incorporating equity into the preservation narrative is not just about redressing the past—it's about reimagining preservation as an inclusive path to sustainability. By broadening the scope of what and whose heritage is preserved, such approaches aim to ensure that preservation serves as a tool for community resilience and continuity, rather than exclusion. This evolution in preservation strategy represents a progressive step towards harmonizing historical integrity with the pressing need for social equity in urban development.

4.4 Redistribution rather than reduction: the complexity of embodied carbon

4.4.1 Production-based emission versus consumer-based emission

The first step towards any equitable sustainability transition is to fully understand the science and logic behind the sustainability measurement. In the case of embodied carbon emission, we want to start from a clearinghouse of different terminologies used in embodied carbon-related discussion.

Production- and consumption-based emissions provide distinct perspectives on a nation's carbon footprint [46]. Production-based emissions account for the greenhouse gases released within a country's borders due to manufacturing and production activities. Consumption-based emissions, in contrast, attribute the greenhouse gases to the goods and services consumed within a country, irrespective of where the emissions occur [47]. This framework underlines the global nature of our consumption patterns, summarized as follows: consumption-based emissions are the sum of all production within a country minus emissions from the production of exported goods plus emissions from the production of imported goods for consumption [48]. The formula for this relationship is straightforward and listed in Box 4.1.

Box 4.1: The relation between consumption- and production-based emissions

$$\text{Consumption-based emission} = \text{Production-based emission} - \text{Exports emission} + \text{Imports emission.}$$

This approach provides a balance sheet where consumption equals domestic production plus imports minus exports. It shifts the focus from the production location to the endpoint of consumption, recognizing the global footprint of consumer habits and the critical role of the supply chain. The disparity between consumption- and production-based emissions highlights the issue of carbon leakage. Carbon leakage occurs when efforts to reduce emissions in one region lead to an increase in another, thereby negating the intended global effort on carbon emission reduction [49, 50]. It is a type of spillover effect or offloading emission burden. It can be positive or negative, expressed as a percentage, and can be greater or less than 100%. Carbon leakage indicator is calculated and linked directly to import and export as shown in Box 4.2.

Box 4.2: Carbon leakage indicator (CLI)

$$\text{CLI} = \text{Trade Intensity} \times \text{Emission Intensity} = \frac{(\text{Imports} + \text{Exports})}{(\text{Turnover} + \text{Imports})} \times \text{Emission Intensity}$$

A 2023 report published on the Energyminute website, illustrated in Figure 4.3, clearly shows the global scale of carbon leakage. For example, Luxembourg has the highest carbon leakage country score, indicating significant emission externalization driven by its consumption demands [51]. It is not

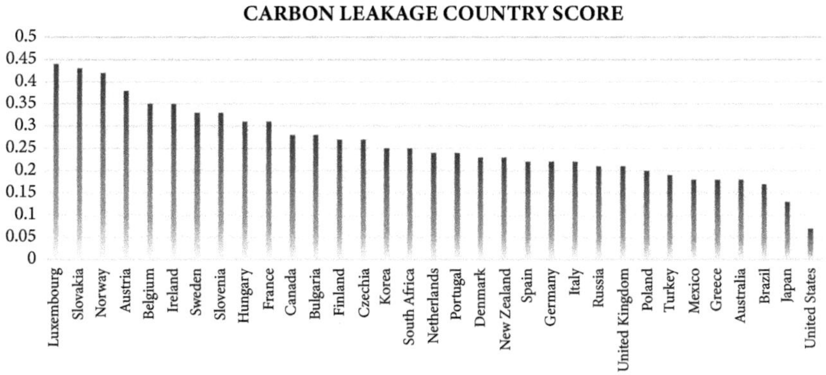

Figure 4.3 Carbon leakage score.

Source: data from Energyminute (2023) *The Perils of Carbon Leakage* https://energyminute.ca/infographics/the-perils-of-carbon-leakage/ [50] [51].

surprising that Luxembourg, with the highest GDP per capita in 2022 [52], also boasts an average household net-adjusted disposable income per capita of $444,773 per year, far exceeding the OECD average [53]. This economic affluence correlates with its high carbon leakage. Luxembourg is not alone; notably, Nordic countries such as Norway, Sweden, and Finland, despite their environmental efforts, show high leakage scores [51]. The report's findings of carbon leakage have sparked considerable discussion on the complexity of unilateral climate actions and their potential to redistribute rather than reduce global carbon emissions merely [54]. What are the results and consequences of carbon leakage? The answer lies in product-based versus consumption-based carbon emission.

Reliable organizations such as the World Bank[1] [55] and the Emission Database for Global Atmospheric Research (EDGAR v8.0) track per-capita emissions data [56, 57], revealing stark differences between production- and consumption-based metrics [58, 59]. Production-based inventories are vital for national emission insights, but consumption-based figures offer a holistic view by considering emissions across the entire production chain attributable to end consumers.

For instance, the United Kingdom saw a reduction of 24% in production-based emissions between 1990 and 2009 post-Kyoto Protocol, but its consumption-based emissions increased by over 1% annually between 1990

[1] World Bank data is sourced directly from the Climate Watch dataset, a reputable and widely used source for greenhouse gas emission data. The World Bank data cover the period from 1990 to 2020 and it is production-based emission data.

and 2008 [48]. Is the UK really making an effort to reduce carbon emissions or just moving their emission to other countries? Table 4.2 juxtaposes consumption- and production-based emissions per capita, highlighting contrasts between nations. Data from 2022 show Sweden and Switzerland, often heralded for sustainability, have low production-based emissions per capita, at 3.69 and 4.11 metric tons, respectively [20]. Yet, this only partially portrays their climate impact. Sweden has reduced emissions by 55.94% since 1990, but its consumption-based emissions are 68.56% greater than its production-based emissions [20]. Switzerland's consumption-based emissions are more than triple its production-based emissions. This means the true carbon footprint of Switzerland and Sweden is far higher than production data alone suggest. The last column of Table 4.2 also shows the extent of the carbon leakage. Again, the question is the following: are these Global North developed countries making an effort to reduce their emissions, or are they merely redistributing carbon emissions caused by their consumption behaviour to other countries? Where did that carbon go?

Conversely, developing countries in the Global South, such as Vietnam, have lower per capita emissions at 3.27 tons, yet 38.06% of their emissions are due to fulfilling consumption demands from other countries [20]. Essentially, the resources and energy Vietnam expends, as well as the environmental toll it bears, are linked to the consumption patterns of individuals in distant lands in developed countries who enjoy the end products without directly facing the environmental costs [20]. It is a complex portrait of affluent consumption patterns that, when scrutinized, reveal an inconvenient truth: *the full environmental cost of our lifestyles is often hidden* [60]. The Global North developed nations are seeing an uptick in consumption-based emissions, suggesting that their economic growth is intricately linked to 'carbon leakage'—the transfer of emissions through international trade [61]. Such discrepancies demonstrate a need for a comprehensive and equitable approach to climate responsibility and challenge us to consider the full scope of our environmental impact in the collective fight against climate change [60].

4.4.2 Value versus cost of embodied carbon

After understanding production- and consumption-based emissions, the second pair of important terminologies are cost and value. The paradox of cost and value refers to the common misunderstanding where individuals conflate

Table 4.2 Consumption- and production-based per capita emission

	Country	Per-capita emission (ton/cap/yr) [20]	Per-capita consumption-based emission (ton/cap/yr) C	Per-capita production-based emission (ton/cap/yr) P	CO_2 emission embedded in global trade (ton CO_2/cap)	
					(C-P)/P (%)	Carbon leakage
Global North	United States	14.44	16.534	15.003	10.21	Moderate
	Japan	8.61	9.829	8.630	13.90	Moderate
	Canada	15.22	13.156	14.341	−8.26	Low
	Germany		9.957	8.273	20.35	Moderate
	France	4.76	6.444	4.808	34.03	High
	United Kingdom	5.00	7.630	5.358	42.42	High
	Netherlands	7.79	9.381	8.513	10.20	Moderate
	Sweden	3.69	6.511	3.863	68.56	High
	Norway	7.62	8.260	7.224	14.35	Moderate
	Finland	6.64	9.338	7.314	27.68	Moderate
	Switzerland	4.11	13.654	4.380	211.73	Extremely High
	Australian	15.12	13.051	14.722	−11.35	Moderate
Global South	China	8.85	7.237	7.903	−8.42	Low
	India	1.91	1.738	1.867	−6.88	Low
	Brazil	2.15	2.156	2.24	−3.73	Low
	Mexico	3.56	3.995	3.62	10.38	Moderate
	Nigeria	0.57	0.617	0.586	2.01	Low
	Indonesia	2.50	2.279	2.227	2.37	Low
	Turkey	5.66	4.977	5.17	−2.74	Low
	Vietnam	3.27	2.329	3.746	−38.06	High
	Thailand	4.06	3.879	3.917	−0.97	Low

A country has high carbon leakage when the (C-P)/P is > 30%. Moderate carbon leakage 10–30%; low carbon leakage is <10%.
Source: data from Our World in Data (2021). Per capita consumption-based CO_2 emission. https://ourworldindata.org/grapher/consumption-co2-per-capita.

the immediate financial cost of an item or action with its overall value, which includes both tangible and intangible benefits accrued over time. Cost is typically the monetary expenditure required to obtain goods or services. This is often the upfront price paid, which can be clearly measured and compared at the point of purchase. On the other hand, value is a more encompassing and subjective measure that includes not only the initial monetary cost but also the benefits derived from the use of the product or service over its life cycle. Value assessments also consider qualitative and quantitative aspects such as satisfaction, utility, environmental impact, and contribution to well-being. The interplay between economic factors and environmental impacts within the realm of embodied carbon involves a nuanced relationship between the cost and value of mitigating or offsetting these emissions over the life cycle of a product or process. The 'cost' of embodied carbon encompasses all expenses incurred from addressing these emissions, ranging from raw material extraction and manufacturing to the operation and eventual deconstruction of a building. Conversely, the 'value' of embodied carbon considers the broader environmental and societal impacts of these emissions [62, 63].

The confusion and misuse of 'cost' and 'value' often stem from a split incentive issue. The initial costs are borne by developers or building owners, while the long-term environmental impacts are distributed among a broader set of stakeholders. For instance, consider the life cycle of embodied carbon in an office building in Washington, D.C., constructed with materials such as aluminium-framed windows and exterior stone panels. Here, the upfront costs of embodied carbon are primarily the responsibility of the developers or investors. Because of the current supply chain of construction products, the cost of aluminium-framed windows does not reflect the actual carbon value and environmental impact of the product, which is artificially low; hence, the developers and investors do not have the incentive to buy locally sourced and made product that could be more expensive measured by financial value. During the operational phase, these costs are shared between the users and the building operators, so they have an immediate need to reduce operational energy, hence the cost.

However, the environmental value, often negative, extends far beyond the immediate stakeholders in Washington, D.C., to affect a larger population and geographic area. This includes the residents near the Bauxite mine in Guinea where the aluminium ore is extracted, the farmers near the quarry in Vermont where the stone is sourced, the fishermen living close to the shipping ports in China where the aluminium windows are shipped from, and urban residents affected by phenomena like the urban heat island effect resulting from the building's construction and operation. Crucially, the true environmental and social value of this office building is often overlooked during the design and planning stages. The building's costs are evaluated in isolation from its broader value, focusing predominantly on immediate financial metrics rather than incorporating a comprehensive assessment of its long-term environmental and societal impacts. This disconnect highlights the need for more integrated approaches in architectural and urban planning that consider both economic and environmental dimensions holistically.

Addressing this paradox involves a shift in perspective from short-term cost minimization to long-term value maximization. Educating consumers and businesses about the broader implications of their purchases and investments, encouraging policies that align costs with true value (such as environmental impact), and developing tools that help quantify and visualize long-term benefits can all play roles in resolving the confusion between cost and value. This shift is crucial for fostering sustainable practices and promoting decisions that consider the welfare of future generations alongside immediate economic benefits.

4.5 The weight of carbon: disparities and challenges in low-income communities

4.5.1 Disproportionate burden on low-income areas

The disproportionate embodied carbon burden on low-income areas is reflected in two key aspects: fuel efficiency and distribution systems and local manufacturers and artisans. Fuel efficiency and distribution systems directly impact embodied carbon by influencing the overall energy consumption in transportation, manufacturing, and construction. Similarly, fostering local manufacturing can reduce reliance on imported goods, thereby lowering embodied carbon emissions throughout life-cycle stages A to C.

Fuel efficiency and distribution systems

In low-income regions, fuel efficiency and distribution systems for materials are often less developed. First, the less efficient electricity mix used during the manufacturing and construction phase highlighted the much higher embodied energy and carbon emissions [64]. With limited infrastructure for efficient transportation, the carbon cost of moving goods can be much higher. This is due to the use of older, less efficient vehicles and machinery, which emit higher levels of pollutants. Furthermore, poor road conditions can exacerbate fuel consumption, increasing emissions. These inefficiencies in the supply chain mean that producing and distributing building materials can have a larger carbon footprint in these areas. In terms of distribution system monitoring and improvement, the traditional data collection methods are time-consuming and unreliable; the need for efficient systems to monitor and assess embodied carbon in construction is crucial for addressing distribution systems and local manufacturing impacts, but such systems are typically lacking in low-income regions [65].

Local manufacturers and artisan

Local manufacturers and artisans in low-income areas may not have access to the technology and materials that allow for low-carbon production. They often work with what is available and affordable, which can mean using more traditional methods that, while sometimes lower-tech, can also be less energy-efficient and more polluting. For example, Hashemi and colleagues studied low-income housing in rural areas in Uganda, and site visits and photographic surveys were carried out to collect information on prevailing construction materials and methods and their related environmental impact [66]. The team found that low-quality, high-waste, energy-intensive production methods, excessive soil extraction, and deforestation are identified as the main environmental damage to current construction methods and materials. The embodied energy is highlighted as the key area that should be addressed to reduce CO_2 emissions of low-income tropical housing. The results indicate that the embodied energy of fired bricks in Uganda is up to 5.7 times more than that of general clay bricks due to the fired bricks' inefficient and energy-wasteful production (using firewood) [66]. Even though concrete has been recognized as an environmentally harmful material, concrete walling is identified as a much more environmentally friendly construction method than brick walling in East African countries compared to carbon-intensive fired bricks [66]. Additionally, reducing the embodied carbon burden is difficult without downstream experts such as quantity surveyors and project managers in the decarbonization effort.

The influence of embodied carbon on low-income areas underscores the need for a more equitable approach to carbon accounting. This approach would recognize the full emissions life cycle and promote investment in infrastructure, renewable energy access, and technological advancements for low-income producers. It would also include supporting local artisans and manufacturers in adopting more sustainable practices, which could reduce the embodied carbon in their products. Overall, reducing the disproportionate burden of embodied carbon on low-income areas requires systemic changes that address the disparities in energy use, production processes, and the global distribution of resources.

4.5.2 Health impact and the socioeconomic divide in carbon burden

The impact of embodied carbon on communities extends well beyond immediate environmental effects, deeply affecting socioeconomic disparities. High embodied carbon infrastructure—such as manufacturing plants and polluting industries—is often disproportionately situated in socioeconomically disadvantaged areas. For example, Figure 4.4 shows the location of cement plants in the United States. Cement is a key embodied carbon emitter, and its production generates significant pollution. Most of those plants are located in low-income communities, as illustrated in Figure 4.4. This increases residents' exposure to pollution and hazardous materials, leading to a spectrum of health risks and perpetuating a cycle of environmental degradation and economic hardship. Higher instances of pollution-related diseases, such as respiratory and cardiovascular conditions, are common in these areas. Proximity to high embodied carbon structures worsens health outcomes, elevates healthcare costs, and reduces productivity, thus exacerbating economic disparity.

A 2019 research led by Dr. Nguyen from the University of Maryland analysing U.S. neighbourhoods revealed that low-income rural communities face higher rates of chronic disease and mortality, highlighting stark health disparities compared to urban areas [67]. An illuminating study by Bell and Keita Ebisu analysed particulate exposures in 215 census tracts between 2000 and 2006, uncovering ethnic and racial disparities in environmental burdens [69]. Hazardous substances such as sulphate—a respiratory irritant—and nickel—a potential carcinogen—were found at significantly higher levels in predominantly Hispanic communities. These communities experienced exposure to nickel and other harmful particulates like nitrate, silicon, and vanadium—linked to cardiovascular and lung diseases—at rates more than 1.5 times higher than those in predominantly white areas. Nickel can be found in stainless steel; silicon can be found in many building products such as sealants, coatings, and waterproofing materials; and vanadium can be found in steel and construction machinery. Similarly, Asian communities were subject to higher levels of hazardous components, including nickel, nitrate, and vanadium. Areas

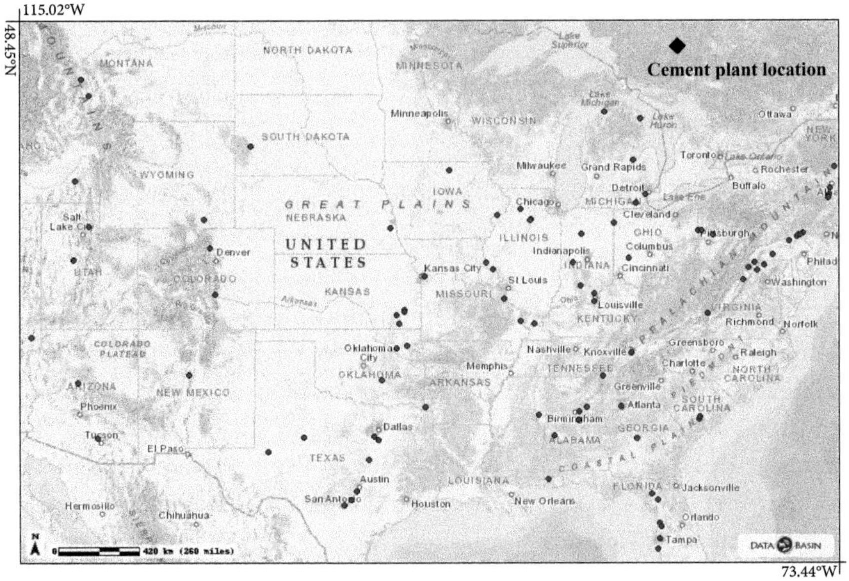

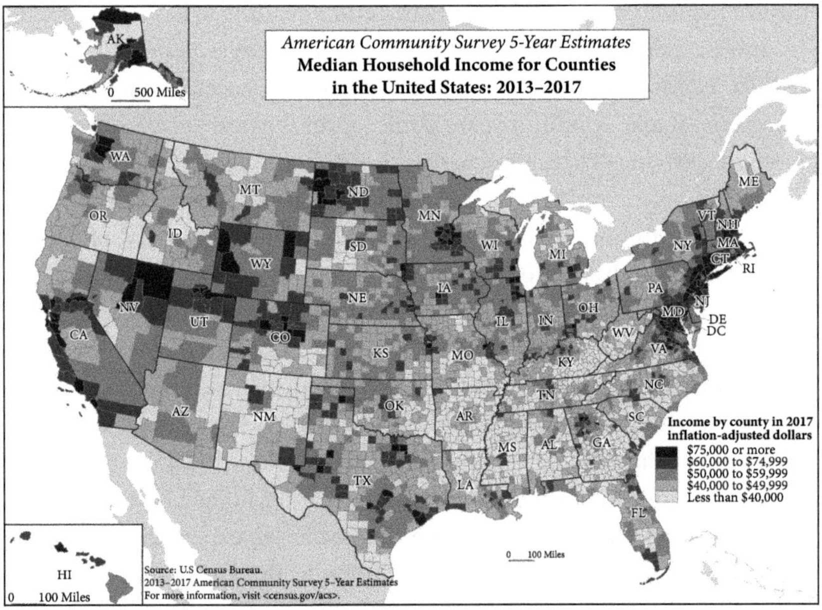

Figure 4.4 Cement plant's location in the United States. (left) The cement plant location.

Source: data from Data Basin (2017). *Cement Plan Locations in US.*; (right) Median household income in the United States: 2013–2017. Source: data from United States Census Bureau (2018). *Median household Income for Counties in the United States: 2013–2017.*

with higher populations of African Americans showed elevated levels of compounds like sulphate and zinc, pointing to a clear pattern of environmental injustice [68].

These insights underscore the urgent need for equitable environmental policies that address the distribution of polluting industries and ensure equal access to clean energy alternatives. Yet, those are just the tip of the iceberg. Next, let's examine the bigger social and economic indication of embodied carbon inequity.

4.6 Conclusion: toward fairer carbon footprints in urban development

As we traverse the intricate terrain of resource distribution and environmental equity, it becomes apparent that the distribution of embodied carbon is not a matter relegated to industry alone but intimately entwined with our societies' socioeconomic fabric. Improving efficiency and being environmentally conscious alone cannot solve the problem, and the industry cannot be the sole party responsible for reducing embodied carbon. This chapter has navigated the disparate realities of urban development, highlighting how marginalized communities in the Global South often bear the brunt of the allocation of environmental burdens and impacts.

Even in Global North developed countries, the contrast between affluent and low-income neighbourhoods is starkly reflected in their access to sustainable resources and clean energy. Less tangible yet equally pervasive disparities in health and well-being find their roots in these unequal distributions. Through the lens of embodied carbon, we observe an environmental challenge and a social dilemma, calling for a paradigm shift in our collective response to urban planning and resource allocation.

Reflecting on the studies presented, it becomes evident that environmental justice requires a holistic approach, integrating clean energy with equitable economic growth and public health considerations. The path forward necessitates an inclusive framework where strategies for reducing embodied carbon are developed with an acute awareness of their social impact. By embracing the lessons from past practices and present insights, policymakers, urban planners, and community leaders are beckoned to craft a future where environmental sustainability is synonymous with social equity. In conclusion, our collective endeavour must not only aim to reduce the carbon footprint of our habitats but to do so in a manner that uplifts every resident, fostering a built environment that is as just as it is sustainable.

References

1. Qi J, Wang Y, Liang S, Li Y, Li Y, Feng C, et al. Primary suppliers driving atmospheric mercury emissions through global supply chains. One Earth. 2019 Oct; 1(2): 254–66.
2. International Trade Administration. Guinea—country commerican guide [Internet]. 2022 [Accessed 14 Apr 2024]. Available from https://www.trade.gov/country-commercial-guides/guinea-mining-and-minerals#:~:text=Guinea%20boasts%2023%20percent%20of,a%20designated%20port%20in%20Kamsar.
3. Dresse A, Nielsen JØ, Fischhendler I. From corporate social responsibility to environmental peacebuilding: The case of bauxite mining in Guinea. Resources Policy. 2021; 74: 102290.
4. Gardner JH, Bell DT. Bauxite mining restoration by Alcoa World Alumina Australia in Western Australia: social, political, historical, and environmental contexts. Restoration Ecology [Internet]. 2007 Dec [accessed 14 Apr 2024]; 15(s4). Available from https://onlinelibrary.wiley.com/doi/10.1111/j.1526-100X.2007.00287.x
5. Ujaczki É, Feigl V, Molnár M, Cusack P, Curtin T, Courtney R, et al. Re-using bauxite residues: benefits beyond (critical raw) material recovery. Journal of Chemical Tech & Biotech. 2018 Sep; 93(9): 2498–510.
6. Chason R, Sharrock C. On frontier of new 'gold rush', quest for coveted EV metals yields misery [Internet]. 2023 [Accessed 14 Apr 2024]. Available from https://www.washingtonpost.com/world/interactive/2023/ev-battery-bauxite-guinea/
7. OEC. Aluminium ore [Internet]. 2022 [Accessed 20 Jun 2024]. Available from https://oec.world/en/profile/hs/aluminium-ore
8. Saadatian S, Freire F, Simões N. Embodied impacts of window systems: a comparative assessment of framing and glazing alternatives. Journal of Building Engineering. 2021 Mar; 35: 102042.
9. ATP Europe. Metal extrusion: what it is and how hot extrusion works [Internet]. [Accessed 20 Jun 2024]. Available from https://atp-europe.com/metal-extrusion-aluminum/
10. Musk AW, De Klerk NH, Beach JR, Fritschi L, Sim MR, Benke G, et al. Respiratory symptoms and lung function in alumina refinery employees. Occup Environ Med. 2000 Apr 1; 57(4): 279–83.
11. Rai S, Wasewar K, Agnihotri A. Treatment of alumina refinery waste (red mud) through neutralization techniques: a review. Waste Manag Res. 2017 Jun; 35(6): 563–80.
12. Chen Y, Li A, Jiang S. Wettability and mechanical properties of red mud–Al2O3 composites. Materials. 2024 Feb 28; 17(5): 1095.
13. OEC. Aluminum doors, windows, frames and thresholds [Internet]. 2022 [Accessed 20 Jun 2024]. Available from https://oec.world/en/profile/hs/aluminium-doors-windows-frames-and-thresholds
14. Böhm ML. Criminal business relationships between commodity regions and industrialized countries: the hard road from raw material to new technology. Journal of White Collar and Corporate Crime. 2020 Jan; 1(1): 34–49.
15. Nenmini Dileep V. Climate change and global health governance in relation to the Global South and the Global North. NAD-Dis [Internet]. 2022 Dec 23 [accessed

11 Apr 2024]; 4(2). Available from https://riviste.unimi.it/index.php/NAD/article/view/19467
16. World Population Review [Internet]. [Accessed 11 Apr 2024]. Available from https://worldpopulationreview.com/country-rankings/energy-consumption-by-country
17. United Nations. Human Development Index (HDI) [Internet]. Human Development Reports. [accessed 2024 Apr 14]. Available from https://hdr.undp.org/data-center/human-development-index#/indicies/HDI
18. Qu J, Maraseni T, Liu L, Zhang Z, Yusaf T. A comparison of household carbon emission patterns of urban and rural China over the 17 year period (1995–2011). Energies. 2015 Sep 22; 8(9): 10537–57.
19. Jackson RB, Ahlström A, Hugelius G, Wang C, Porporato A, Ramaswami A, et al. Human well-being and per capita energy use. Ecosphere. 2022 Apr; 13(4): e3978.
20. European Commission. GHG emissions of all world countries (2023 report) [Internet]. 2023 [Accessed 2024 Apr 21]. Available from https://edgar.jrc.ec.europa.eu/report_2023
21. International Energy Agency. Solar PV manufacturing capacity by country and region, 2021 [Internet]. 2022 [Accessed 17 Apr 2024]. Available from https://www.iea.org/data-and-statistics/charts/solar-pv-manufacturing-capacity-by-country-and-region-2021
22. Wiliani N, Rahman T, Sani A, Ramli S. Statistical lcharacteristics for identification defect of solar panel with naive Bayes [Internet]. 2019 [Accessed 17 Apr 2024]. Available from https://osf.io/uyfne
23. Sharma A, Mahajan P, Garg R. End-of-life solar photovoltaic panel waste management in India: forecasting and environmental impact assessment. Int J Environ Sci Technol. 2024 Jan; 21(2): 1961–80.
24. Buitrago E, Novello AM, Meyer T. Third-generation solar cells: toxicity and risk of exposure. Helvetica Chimica Acta. 2020 Sep; 103(9): e2000074.
25. Brock A, Sovacool BK, Hook A. Volatile photovoltaics: green industrialization, sacrifice zones, and the political ecology of solar energy in Germany. Annals of the American Association of Geographers. 2021; 111(6): 1756–78.
26. Sovacool BK. Who are the victims of low-carbon transitions? Towards a political ecology of climate change mitigation. Energy Research & Social Science. 2021; 73: 101916.
27. Mulvaney D. Are green jobs just jobs? Cadmium narratives in the life cycle of photovoltaics. Geoforum. 2014; 54: 178–86.
28. Cyrs WD, Avens HJ, Capshaw ZA, Kingsbury RA, Sahmel J, Tvermoes BE. Landfill waste and recycling: Use of a screening-level risk assessment tool for end-of-life cadmium telluride (CdTe) thin-film photovoltaic (PV) panels. Energy Policy. 2014; 68: 524–33.
29. Sinha P, Trumbull VL, Kaczmar SW, Johnson KA. Evaluation of potential health and environmental impacts from end-of-life disposal of photovoltaics. In: Photovoltaics: Synthesis, Applications and Emerging Technologies. Nova Publishers; 2014. pp. 37–52.
30. Genchi G, Sinicropi MS, Lauria G, Carocci A, Catalano A. The effects of cadmium toxicity. IJERPH. 2020 May 26; 17(11): 3782.

31. Cross J, Murray D. The afterlives of solar power: waste and repair off the grid in Kenya. Energy Research & Social Science. 2018; 44: 100–9.
32. Salim HK, Stewart RA, Sahin O, Dudley M. Drivers, barriers and enablers to end-of-life management of solar photovoltaic and battery energy storage systems: a systematic literature review. Journal of Cleaner Production. 2019; 211: 537–54.
33. Sovacool BK, Hook A, Martiskainen M, Brock A, Turnheim B. The decarbonisation divide: contextualizing landscapes of low-carbon exploitation and toxicity in Africa. Global Environmental Change. 2020; 60: 102028.
34. Karlen C, Pagani A, Binder CR. Obstacles and opportunities for reducing dwelling size to shrink the environmental footprint of housing: tenants' residential preferences and housing choice. J Hous and the Built Environ. 2022 Sep; 37(3): 1367–408.
35. Hu M. Optimal renovation strategies for education buildings—a novel BIM–BPM–BEM framework. Sustainability. 2018; 10(9): 3287.
36. Hu M. Balance between energy conservation and environmental impact: life-cycle energy analysis and life-cycle environmental impact analysis. Energy and Buildings. 2017; 140: 131–9.
37. Bullen PA, Love PED. Adaptive reuse of heritage buildings. Structural **Survey**. 2011 Nov 8; 29(5): 411–21.
38. Cucco P, Maselli G, Nesticò A, Ribera F. An evaluation model for adaptive reuse of cultural heritage in accordance with 2030 SDGs and European Quality Principles. Journal of Cultural Heritage. 2023 Jan; 59: 202–16.
39. Cheong C. Preservation at a crossroads: the need for equity preservation. Change over Time. 2021; 10(1): 66–83.
40. Bertolet D. When historic preservation clashes with housing affordability [Internet]. 2017 [Accessed 17 Apr 2024]. Available from https://www.sightline.org/2017/12/19/when-historic-preservation-clashes-with-housing-affordability/
41. Cortright J. Historic preservation: NIMBYism for the rich? [Internet]. 2017 [Accessed 17 Apr 2024]. Available from https://www.strongtowns.org/journal/2017/6/13/historic-preservation-nimbyism-for-the-rich
42. McCabe BJ, Ellen IG. Does preservation accelerate neighborhood change? Examining the impact of historic preservation in New York City. Journal of the American Planning Association. 2016 Apr 2; 82(2): 134–46.
43. Arkaraprasertkul N. Gentrifying heritage: how historic preservation drives gentrification in urban Shanghai. International Journal of Heritage Studies. 2019; 25(9): 882–96.
44. Minner J. Revealing synergies, tensions, and silences between preservation and planning. Journal of the American Planning Association. 2016 Apr 2; 82(2): 72–87.
45. Lephew J, Gajewski J. Historic preservation and the housing crisis: a 'two birds, one stone' solution [Internet]. 2022 [Accessed 17 Apr 2024]. Available from https://www.globest.com/2022/02/28/historic-preservation-and-the-housing-crisis-a-two-birds-one-stone-solution/?slreturn=20240317111311
46. Jakob M, Steckel JC, Edenhofer O. Consumption-versus production-based emission policies. Annu Rev Resour Econ. 2014; 6(1): 297–318.
47. Peters GP. From production-based to consumption-based national emission inventories. Ecological Economics. 2008; 65(1): 13–23.

48. Barrett J, Peters G, Wiedmann T, Scott K, Lenzen M, Roelich K, et al. Consumption-based GHG emission accounting: a UK case study. Climate Policy. 2013 Jul; 13(4): 451–70.
49. Yu B, Zhao Q, Wei YM. Review of carbon leakage under regionally differentiated climate policies. Science of the Total Environment. 2021; 782: 146765.
50. European Parliament. Carbon leakage: preventing firms from avoiding emissions rules [Internet]. [Accessed 26 Apr 2024]. Available from https://www.europarl.europa.eu/topics/en/article/20210303STO99110/carbon-leakage-preventing-firms-from-avoiding-emissions-rules
51. Foyer A. The perils of carbon leakage [Internet]. 2023 [Accessed 2024 Apr 26]. Available from https://energyminute.ca/infographics/the-perils-of-carbon-leakage/
52. Worldometer. GDP per capita [Internet]. [Accessed 19 Jun 2024]. Available from https://www.worldometers.info/gdp/gdp-per-capita/
53. Luxembourg. OECD Better Life Index [Internet]. [Accessed 19 Jun 2024]. Available from https://www.oecdbetterlifeindex.org/countries/luxembourg/
54. Eckersley R. The politics of carbon leakage and the fairness of border measures. Ethics & International Affairs. 2010; 24(4): 367–93.
55. Climatewatch. Climate watch GHG emission module [Internet]. [cited 2024 Apr 21]. Available from https://www.climatewatchdata.org/about/faq/ghg
56. The International Energy Agency (IEA). Greenhouse gas emissions from Energy Data Explorer [Internet]. 2023 [Accessed 21 Apr 2024]. Available from https://www.iea.org/data-and-statistics/data-tools/greenhouse-gas-emissions-from-energy-data-explorer
57. European Commission. EDGAR—Emissions Database for Global Atmospheric Research [Internet]. [Accessed 21 Apr 2024]. Available from https://edgar.jrc.ec.europa.eu/dataset_ghg80
58. Mi Z, Zhang Y, Guan D, Shan Y, Liu Z, Cong R, et al. Consumption-based emission accounting for Chinese cities. Applied Energy. 2016 Dec; 184: 1073–81.
59. Afionis S, Sakai M, Scott K, Barrett J, Gouldson A. Consumption-based carbon accounting: does it have a future? WIREs Climate Change. 2017 Jan; 8(1): e438.
60. Cao Y, Qu S, Zheng H, Meng J, Mi Z, Chen W, et al. Allocating China's CO_2 emissions based on economic welfare gains from environmental externalities. Environ Sci Technol. 2023 May 23; 57(20): 7709–20.
61. Grubb M, Jordan ND, Hertwich E, Neuhoff K, Das K, Bandyopadhyay KR, et al. Carbon leakage, consumption, and trade. Annual Review of Environment and Resources. 2022; 47: 753–95.
62. Hammond GP, Jones CI. Embodied energy and carbon in construction materials. Proceedings of the Institution of Civil Engineers-Energy. 2008; 161(2): 87–98.
63. Ritter H, Schopf M. Unilateral climate policy: harmful or even disastrous? Environ Resource Econ. 2014 May; 58(1): 155–78.
64. Turconi R, Boldrin A, Astrup T. Life cycle assessment (LCA) of electricity generation technologies: Overview, comparability and limitations. Renewable and Sustainable Energy Reviews. 2013 Dec; 28: 555–65.
65. Xu J, Pan W, Teng Y, Zhang Y, Zhang Q. Internet of things (IoT)—integrated embodied carbon assessment and monitoring of prefabricated buildings. IOP Conf Ser: Earth Environ Sci. 2022 Nov 1; 1101(2): 022031.

66. Hashemi A, Cruickshank H, Cheshmehzangi A. Environmental impacts and embodied energy of construction methods and materials in low-income tropical housing. Sustainability. 2015 Jun 18; 7(6): 7866–83.
67. Nguyen QC, Khanna S, Dwivedi P, Huang D, Huang Y, Tasdizen T, et al. Using Google Street View to examine associations between built environment characteristics and U.S. health outcomes. Preventive Medicine Reports. 2019 Jun; 14: 100859.
68. Bell ML, Dominici F, Ebisu K, Zeger SL, Samet JM. Spatial and temporal variation in PM2. 5 chemical composition in the United States for health effects studies. Environmental Health Perspectives. 2007; 115(7): 989–95.
69. Bell ML, Ebisu K. Environmental inequality in exposures to airborne particulate matter components in the United States. Environmental Health Perspectives. 2012; 120(12): 1699–704.

Chapter 5
Global North bias in sustainability

5.1 What is Global North bias?

5.1.1 Definition of Global North and South

The Global North and Global South are not geographical locations but describe countries which are in a privileged societal, political, and economic position (Global North) or a disadvantaged position globally (Global South). We might be able to trace the term Global North and South back to a report produced by the independent Commission on International Development Issues, chaired by former West German Chancellor Willy Brandt in 1980 [1]. This report provided an understanding of drastic differences in the economic development of the Global North and Global South. Brant also proposed the definition of Global North and South based on gross domestic production (GDP) per capita, which can be visualized as the Brandt line, as illustrated in Figure 5.1. Since then, the categories Global South and Global North have also been widely applied in urban (sustainability) studies in general due to the benefits compared to other categories such as developed/developing countries [2, 3].

Built upon Brant's concept, the United Nations Conference on Trade and Development (UNCTAD) categorizes the Global North as primarily comprising affluent countries. This group includes North America (the United States and Canada), Western Europe (the United Kingdom and European Union nations), Russia, Australia, New Zealand, and selected East Asian countries like Japan and South Korea, as well as Israel and Singapore. In contrast, the Global South encompasses Africa, Latin America and the Caribbean, the Pacific Islands, and the entirety of Asia. Notable within this group are rapidly growing economies such as China, India, Brazil, and Indonesia, along with significant emerging markets like Nigeria and Mexico. This classification helps in understanding the economic disparities and developmental challenges that distinctly affect these groups on the global stage [4]. Since then, it has been widely used in studying various global issues, such as sustainability, climate change, and carbon trade.

Embodied Carbon. Ming Hu, Oxford University Press. © Ming Hu (2025). DOI: 10.1093/9780198944485.003.0005

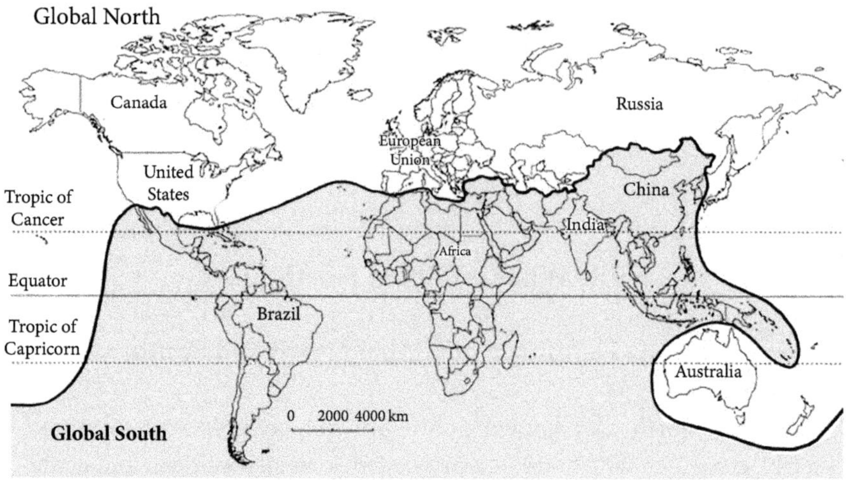

Figure 5.1 Brandt line.

Reproduced from Jovan.gec via Wikimedia Commons, (2015), https://en.wikipedia.org/wiki/Brandt_Report#/media/File:The_Brandt_Line.png under a Creative Commons Attribution-ShareAlike 4.0 International (CC BY-SA 4.0)

5.1.2 Understanding Global North bias

The concept of 'Global North bias in sustainability' highlights an imbalance in how sustainability efforts are prioritized and addressed, often reflecting the perspectives and needs of the Global North while marginalizing those of the Global South. This bias permeates various areas of sustainability research and practice. Notable examples include the 2021 International Energy Agency (IEA) report, 'Zero by 2050: A Roadmap for the Global Energy Sector', and recent discussions and documents from the Intergovernmental Panel on Climate Change (IPCC). These examples underscore how the urgent and necessary actions for sustainability can sometimes overshadow the unique challenges and voices from the Global South [5].

The IEA's report called for an immediate halt to new investments in unabated coal plants, eliminating the least efficient coal plants by 2030, and converting all operational coal plants by 2040. However, this directive faced resistance from numerous developing and emerging economies in the Global South, particularly China. An official from a Chinese state-affiliated research body expressed concerns about meeting the growing energy demand solely with non-fossil fuel sources, suggesting the necessity of fossil fuels to bridge the potential supply gap [6]. China, with the world's fourth-largest coal reserves, produces around 90% of the coal it consumes, compared to importing 70% of its oil and 50% of its gas [7]. Over the past forty years, China's economy has grown at an average annual rate of 9.5%, largely propelled by coal [7]. Meanwhile, China has made significant efforts to transition to green energy, with its installed solar and wind energy capacity accounting for 35–40% of the global total as of 2021 [8]. According to the IEA, China is expected to reach 1,200 gigawatts (GW) of total wind and solar capacity by 2026, four years ahead of its current target of 2030 [9], while the U.S. expects to have 306 GW of wind and solar capacity by 2026, according to U.S. Energy Information Administration [10]. These data indicate that China has a much more aggressive plan to transition to green energy, with four times the capacity of the United States. Countries like China in the Global South argue that because developing countries need to grow to improve residents' quality of life to levels comparable with the Global North, their energy consumption will inevitably increase. A large portion of the energy use in China is caused by the consumer needs of the Global North developed countries. Additionally, these countries have already made significant efforts to transition to green energy, more aggressively than many Global North countries. Therefore, one-size-fits-all policies on coal are viewed as unsuitable and unfair to the Global South.

Geopolitics also plays a crucial role in China's energy policy. The country's reliance on coal over imported oil and gas stems from a 'deep-rooted fear' of isolation due to historical sanctions [7]. After the Communist Party took over in 1949, China faced simultaneous supply and technical restrictions from the U.S. and the former Soviet Union. This experience underscored the importance of self-reliance for key energy supplies to the central government. Maintaining energy independence is seen as essential for political and social stability [7]. This perspective highlights the impracticality and inequity of applying a uniform timeline to both developing and developed nations, given their different historical contributions to cumulative emissions and diverse developmental and energy consumption paths. The neglect of historical cumulative emissions is not an isolated incident; it has increasingly been a muted topic in past IPCC gatherings, as various scholars have noted [4, 11,12].

Furthermore, the focus of IPCC discourses and reports appears to be shifting from mitigation to adaptation [13]. The rationale behind this shift seems transparent. Mitigation requires collective global action rooted in historical accountability, while adaptation revolves around localized actions, often powered by investments in green technology. The affluent nations of the Global North, equipped with resources and choice, can afford the costs associated with drastic emission reductions for the greater good—or opt to bolster their own infrastructures against the impact of extreme weather. However, the preference has been leaning towards adaptation, effectively sidelining the critical need for immediate greenhouse gas reduction. This choice, less consequential for the wealthy, is poised to impose a heavy toll on the developing nations of the Global South, which find themselves without the means to either effectively mitigate or adapt [4].

The Global North bias is mirrored in the way value chain actors from these regions employ standards and certifications to assert the sustainability of products and practices in the Global South [14]. Additionally, leadership in research on sustainable energy transitions and infrastructure often resides with the Global North, potentially neglecting the unique priorities and perspectives of the Global South [15]. This inclination is similarly noticeable in the crafting of sustainable development policies, where the specific circumstances and necessities of the Global South may not be adequately acknowledged or addressed [16].

A poignant instance of such bias was detailed in an article by Dr Harini Nagendra, published in *Nature*, titled 'The Global South is Rich in Sustainability Lesson That Students Deserve to Hear' [3]. The imposition of a ban on livestock grazing precipitated ecological deterioration in India's Keoladeo National Park, a biodiverse wetland in Rajasthan noted for its avian population. The policy, modelled on American conservation practices aimed at safeguarding supposedly untouched landscapes from disturbance, led to the expulsion of cattle from Keoladeo in 1982 [3, 17]. Post-ban, the park's ecosystem entered a period of gradual decline. The absence of grazing cattle allowed canals to become overgrown with weeds and grasses, a diet they previously kept in check, resulting in a stark reduction in the diversity of birds and other wildlife. This is just one such example illustrating that superimposing so-called 'sustainable practice' from the Global North on the Global South without understanding the local conditions and needs is not only impractical but also damaging.

The pervasive influence of the Global North in dictating sustainability practices raises fundamental questions about fairness and equity, particularly when such policies fail to account for the diverse ecological and economic realities across the globe. The case of Keoladeo National Park exemplifies how well-intentioned policies can lead to unintended ecological harm when they ignore local environmental dynamics and community needs. As we transition into discussions on environmental and climate justice, it becomes clear that achieving true sustainability requires addressing the inequalities embedded in global environmental policies. These concepts of justice not only challenge us to consider the rights and voices of marginalized communities but also compel us to rethink the structures of power and decision-making that shape our global response to environmental challenges. Section 5.2 explores how integrating environmental justice and climate justice into sustainability practices can offer a more equitable and effective framework for addressing global environmental issues, ensuring that all communities, especially those most vulnerable, have a say in the policies that affect their lives and environments.

5.2 Climate justice: Global North and South dynamics

5.2.1 The inception and evolution of environmental justice

The concept of environmental justice gained prominence through the pioneering work of scholars Dr. Paul Mohai, Dr. David Pellow, and Dr. J. Timmons Roberts in their seminal piece, 'Environmental Justice'. This framework, which first emerged in the United States during the 1970s, arose in response to the unequal distribution of environmental burdens across socioeconomically disadvantaged communities [18]. It catapulted into the national spotlight in 1982 when activists halted the dumping of 120 million pounds of polychlorinated biphenyls (PCBs) contaminated soil in a predominantly African American county in North Carolina [19]. This movement laid the groundwork for the U.S. Environmental Protection Agency's definition of environmental justice, which emphasizes the fair treatment and meaningful involvement of all people regardless of race, colour, national origin, or income in environmental policy development and enforcement [20, 21]. The movement primarily focuses on issues like the siting of hazardous facilities in marginalized areas and efforts to remediate pollution impacts on affected communities [22].

5.2.2 Global environmental justice and carbon debt dynamics

The discourse on environmental justice has expanded to encompass global perspectives, focusing notably on the disparities in carbon emissions and trade between the developed Global North and the developing Global South [23]. Typically, developed regions import carbon-intensive, low-price products from less developed areas, effectively offloading the environmental burden of carbon emissions back onto these poorer regions. This transfer of burden is captured under the concept of carbon debt, also known as ecological debt, which underscores the economic and environmental imbalances in international trades [24].

While sharing roots with environmental justice, the concept of carbon debt ventures further into the realm of climate justice. Environmental justice primarily addresses the local distribution of environmental harms. In contrast, climate justice adopts a broader lens, incorporating global ethical, moral, and human rights dimensions related to climate change [19]. It emphasizes that both the origins and impacts of climate change disproportionately affect the world's most vulnerable populations, highlighting a critical area for global policy intervention [25, 26].

A pivotal study conducted by Dr Damon Matthews and published in *Nature Climate Change* (2016) has quantified carbon debt as the cumulative disparity between a nation's actual CO_2 emissions and its equitable share, calculated based on population [27]. This assessment has revealed a significant carbon debt owed by the Global North, aggravated by historical emissions and ongoing industrial activities [27]. Meanwhile, many Global South countries are credited with carbon credits for their relatively lower contributions to global emissions. According to the formula (Box 5.1, Equation 1), the Global North collectively owes approximately 589.83 Gt of CO_2, with the United States alone responsible for 348.88 Gt, followed by the European Union and the Russian Federation. Conversely, the Global South holds a carbon credit of 555.13 Gt CO_2, with India and China being notable holders [4].

Box 5.1: Carbon share

$$\text{Allotted Share} = \frac{\text{Country Population}}{\text{Global Population}} \times \textit{World Emission} \tag{5.1}$$

This calculation vividly illustrates (Figure 5.2) the considerable carbon debt accrued by the Global North, suggesting a potential shift towards a degrowth model for mitigating emissions and rectifying imbalances with the Global South.

However, the prevailing narrative often portrays the Global North as a global sustainability leader, continuing to support the Global South in reducing carbon emissions. This perception persists despite the substantial carbon debt, highlighting a disconnect between the reality of carbon debt and the

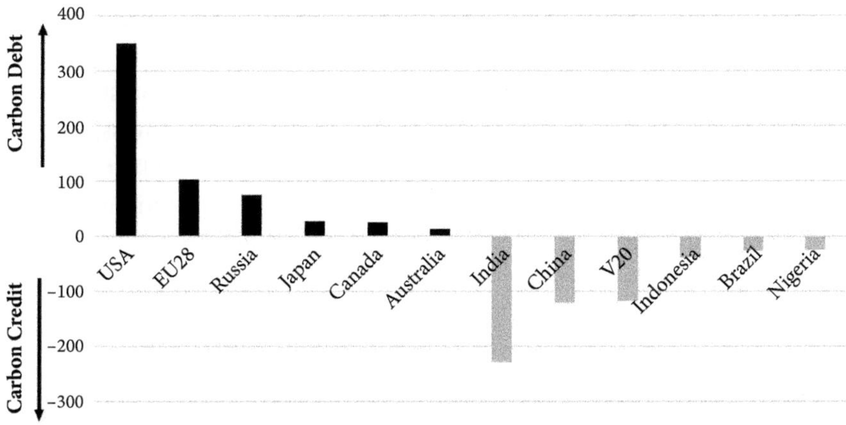

Figure 5.2 Historical cumulative carbon debts and credits from 1850 to 2019 (in Gt. CO_2).

Adapted from Singh and Bajwa, 'Who Is Responsible For The Climate Crisis? A Perspective from The Global South. International Critical Thought' (2023), Taylor & Francis, doi:10.1080/21,598,282.2023.2256474

ongoing global climate policy discussions. This discrepancy calls for a thorough reassessment of international climate strategies, emphasizing equity, reparations, and the sustainable development of all nations.

5.2.3 Reassessing the narrative of global climate cooperation

The portrait of the Global North as a benevolent partner assisting the Global South in mitigating carbon emissions is evident in the political dominance of Global North countries in global climate discussions [4]. However, it is increasingly contested by empirical studies. Historical responsibilities towards climate change are frequently overlooked, with blame disproportionately cast on emerging economies such as India and China. These nations are often criticized for their rising emissions, attributed to loose regulations and reliance on coal and fossil fuels. This narrative often forgets that developed countries in the Global North also went through similar energy-intensive and carbon-heavy periods to achieve their current high living standards. Even today, they maintain lifestyles that are significantly carbon-intensive compared to those who live in the Global South, as shown in Chapter 4's Table 4.2. This Global North-dominated narrative creates a roadblock to genuine progress in addressing climate change.

To illustrate the importance of carbon justice, consider Bhutan as a poignant example. Nestled in the Himalayas, Bhutan has maintained its carbon-negative status for over five decades through dedicated efforts in reforestation and renewable energy development. [28] In a powerful 2016 address, Bhutan's Prime Minister Tshering Tobgay vividly described the existential threats posed by climate change to his country, underscoring the disproportionate impact on nations that have contributed least to global warming. He highlighted the devastating effects of melting glaciers, which have formed new lakes prone to causing catastrophic floods.

> Climate change is affecting my country. Our glaciers are melting, causing flash floods and landslides, which in turn are causing disaster and widespread destruction in our country. I was at that lake recently. It's stunning. That's how it looked 10 years ago, and that's how it looked 20 years ago. Just 20 years ago, that lake didn't exist. It was a solid glacier. A few years ago, a similar lake breached its dams and wreaked havoc in the valleys below. That destruction was caused by one glacier lake. We have 2,700 of them to contend with. The point is this: my country and my people have done nothing to contribute to global warming, but we are already bearing the brunt of its consequences. And for a small, poor country, one that is landlocked and mountainous, it is very difficult. But we are not going to sit on our hands doing nothing. We will fight climate change. That's why we have promised to remain carbon neutral.

Bhutan's message was clear: despite Bhutan's minimal contributions to global warming, it faces severe consequences and remains committed to staying carbon neutral. This example underscores the need for a fairer approach to climate responsibility that acknowledges the historical emissions of Global North developed nations and supports the sustainable development of those most affected by climate change.

Recent research led by Dr. Jason Hickel, published in *Nature Sustainability*, highlights a critical issue: even under the most optimistic IPCC scenarios, the Global North is on track to exceed its equitable share of the carbon budget by a factor of three, significantly encroaching upon the share allotted to the Global South. A 2023 study published in *Nature Sustainability* estimates that by 2050, the North could owe up to $192 trillion in compensation for these disparities, which breaks down to about $940 per capita annually in the South [29]. Such findings underscore the necessity of 'carbon justice', a concept that addresses these imbalances by advocating for fair and equitable distribution of carbon responsibilities and reparations for historical overemissions.

This growing body of evidence necessitates a thorough reevaluation of global climate policies. It calls for the adoption of strategies that not only comprehensively tackle carbon emissions but also rectify the profound economic disparities and injustices ingrained in current global trade and environmental practices. The question is, what factors contribute to persistent bias and misperception? To move forward, the first step is to reveal those root causes. Section 5.3 identifies two critical factors that contribute to and drive these dynamics and carbon-climate injustice: data fallacy and disregarding local knowledge.

5.3 Cause for Global North–South misperception

5.3.1 Data fallacy in climate accountability

Scholars and activists, particularly those from the Global South, have consistently highlighted the imbalanced influence exerted by the Global North on the climate change discourse. A glaring example of this imbalance is the selective use of data, which can distort logic and perpetuate incorrect assumptions and biases [5]. This issue is especially evident in the methodologies used for accounting for carbon emissions.

As elucidated earlier in Chapter 4, both production- and consumption-based carbon emissions provide critical insights into a country's carbon contributions—the former tracks domestic direct emissions, while the latter accounts for carbon embodied in consumed traded goods. Ideally, discussions should balance both metrics. However, there is a prevalent tendency to focus on production-based statistics, overshadowing consumption-based data, with national aggregate-level carbon figures taking precedence over per-capita data. This approach unfairly places a disproportionate burden of emissions on large developing and emerging economies. Economist Dieter Helm points out in his book *Legacy: How to Build the Sustainable Economy* that this distortion can lead to misdirected responsibilities [30]:

> Polluters should pay for the pollution they cause. This is a secondary principle of the sustainable economy, alongside the precautionary principle. Ultimately, it is you and me who buy the stuff, and hence we are the ultimate polluters. Since we are currently not paying the full costs of that pollution, it is another reason why we are living beyond our environmental means, and hence implies one more notch down in consumption.

Furthermore, the 'global emissions' section on the U.S. Environmental Protection Agency (EPA) website predominantly showcases production-based and population-aggregated emissions, categorizing countries like China, the United States, and India as the top three emitters, without a critical discussion of the inherent biases of this method [31]. This pattern persists in other influential reports, such as 'Net Zero by 2050: A Roadmap for the Global Energy Sector', which predominantly foreground production-based statistics [32]. Although these reports occasionally acknowledge the importance of consumption-based data, such nuances are often relegated to the margins of the discourse, maintaining a focus on production.

When emissions are analysed through the lens of consumption and on a per-capita basis, a dramatically different narrative emerges. For instance, Luxembourg displays one of the highest carbon footprints per individual, far outstripping the footprints of countries like Madagascar and Uganda. Similarly, the U.S. footprint per capita far exceeds those of many developing countries [5]. This oversight diverts attention from a critical truth: Global North affluent countries bear significant moral responsibility for current emissions and should thus face accountability for their contributions to global burdens.

Another critical oversight in the climate discourse is the frequent skewing of historical cumulative carbon emissions in favour of recent data. Historical or cumulative emissions are essential for illustrating climate injustice, as CO_2 persists in the atmosphere for centuries. Yet, international institutions and wealthy country blocs often downplay the discussion of historical emissions, skewing the allocation of climate change responsibility [4]. This tendency can be seen as a form of ecological bias, where the developed Global North, in concert with international bodies, directs the focus towards emerging economies [5]. Recent shifts in how historical cumulative data are illustrated further exemplify the intentional manipulation of this data. For example, re-evaluations by Carbon Brief extended historical emissions data from 1850 to 2023, significantly altering the perceived contributions of various countries. Analysing historical emissions per capita, as shown in Figure 5.3a, reveals that when emissions are calculated on a per-capita basis from 1850 to 1950 (historical year), the rankings change dramatically, highlighting the disproportionate contributions of countries like the U.S. and the EU. However, this is not the picture the Global North focuses on. Instead, Figure 5.3b, which counts emissions until 2022, is commonly presented, reinforcing a narrative that shifts the burden of responsibility away from historically high emitters.

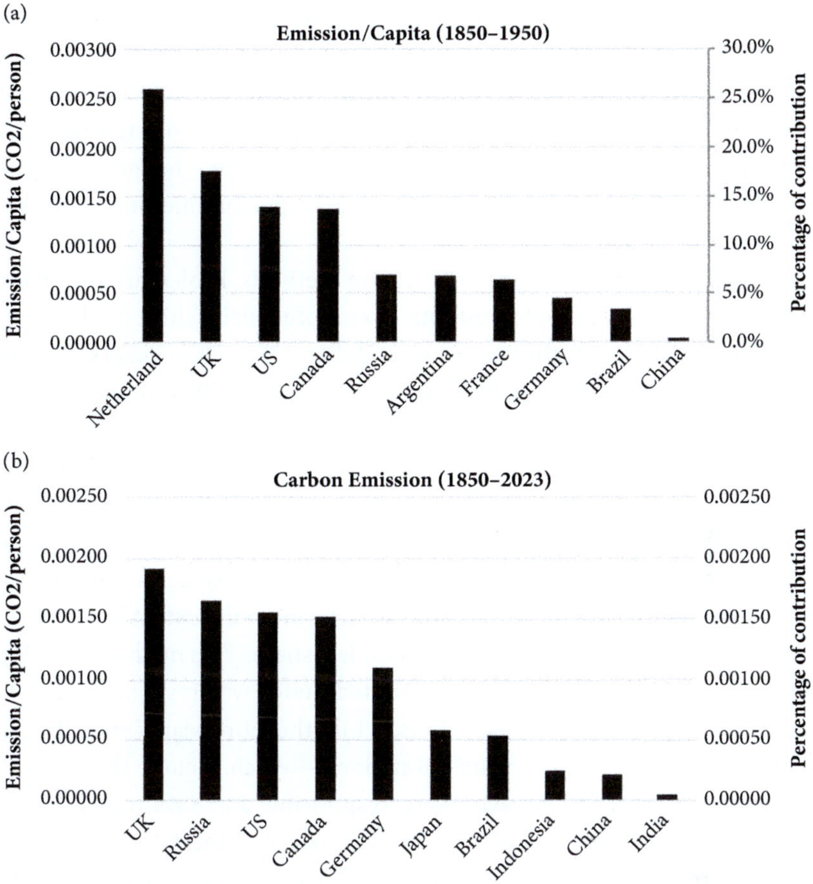

Figure 5.3 (a) Historical cumulative CO_2 emissions (1850–1950). (b) Historical cumulative CO_2 emissions (1850–2023).

Source: Data from Carbon Brief (2023) *Revealed: How colonial rule radically shifts historical responsibility for climate change* [33].

Acknowledging historical responsibility does not exempt the Global South developing world from addressing today's climate threat. Indeed, while historical emissions are crucial, the urgent nature of the climate crisis and the need for global collaboration transcends any single narrative. It's about striking a balance between acknowledging the past and taking collective action in the present, ensuring that all nations contribute fairly to the global efforts to mitigate and adapt to climate change. This approach supports a more equitable assessment of historical emissions and encourages a more nuanced understanding of each country's current responsibilities and capabilities [34].

5.3.2 Towards a more equitable understanding

This selective framing and methodological tweaking of data challenge our understanding of true climate responsibility and underscore the need for a more equitable approach to addressing global climate change. Manipulating emission data and downplaying historical responsibilities not only distort public perception but also policy directions.

In a notable contrast to more mainstream methods, Professor Jason Hickel, an economic anthropologist, adopts a comprehensive historical perspective in his analysis of climate accountability [35]. Dr. Hickel's approach involves extending the historical period back to 1850 and incorporating both territorial and consumption-based emissions. This dual consideration allows for a fuller accounting of each country's contributions to global emissions, particularly emphasizing the impact of consumption in developed nations.

Dr. Hickel calculates national shares of the safe global carbon budget subtracted from actual historical emissions to determine the extent to which each country has overshot or stayed within their fair share. This methodology highlights a stark reality: as of 2015, the United States was responsible for 40% of the global excess in CO_2 emissions, while the European Union (EU-28) accounted for 29%. Together, the G8 nations—which include the USA, EU-28, Russia, Japan, and Canada—were responsible for 85% of these excess emissions. Impressively, Hickel's analysis shows that the Global North as a whole was responsible for 92% of excess global emissions, starkly contrasting with most countries in the Global South, which largely remained within their carbon boundaries [35]. By broadening the timeframe and incorporating consumption-based metrics, Hickel provides a critical corrective to prevalent narratives that disproportionately implicate emerging economies in the climate crisis.

Dr. Hickel's analysis calls for re-evaluating global climate policies, suggesting that the developed countries must make more significant adjustments to address their historical overconsumption. This perspective not only challenges the conventional data presentation but also pushes for a deeper reconsideration of fairness and equity in global climate agreements. To address these data fallacies effectively, there is a crucial need to advocate for transparency in data presentation and to push for methodologies that reflect both historical responsibilities and current capabilities.

5.3.3 Embracing local wisdom: bridging sustainable development and local knowledge

Besides the data, the prejudice towards local knowledge, especially vernacular knowledge, created huge barriers and obstacles for Global South countries to become equal players to raise concern. The Global North 'knows how' attitude is detrimental to real sustainability. In recent years, the Global North's approach to sustainable development has increasingly come under scrutiny for its lack of sensitivity to local contexts. This critique has spurred a burgeoning interest in tapping into local, traditional, and vernacular knowledge to craft innovative, sustainable solutions.

A particularly compelling example of this shift is Diébédo Francis Kéré, a Burkinabé-German architect whose work has consistently fused sustainability with local knowledge. In 2022, Kéré became the first African to win the prestigious Pritzker Architecture Prize, underscoring his pioneering approach. Kéré's design philosophy often challenges the prevailing practices where inexperienced architects impose modern building techniques and materials in Global South developing countries, leading to unsustainable outcomes. In one of his notable projects, he confronted the oppressive heat of Burkina Faso not with air conditioning, which was impractical due to high electricity costs and lack of infrastructure and maintenance, but through innovative use of local materials and passive cooling techniques. As shown in Figure 5.4, the classrooms he designed utilized locally sourced clay and incorporated design elements like passive ventilation, underground cooling systems, double-skin light-weight roofs, and strategic vegetation, proving that 'combining traditional and contemporary materials and techniques leads to affordable and sustainable solutions' [36].

However, Kéré's approach initially met strong resistance, sometimes even from the local community. When he proposed building a clay school in his hometown Gando (Figure 5.4) after his studies in Germany, the community was initially proud but soon became disdainful when they learned of the building materials. Clay and laterite, despite their practical benefits, were seen as symbols of poverty, not progress, regarded as 'poor man's materials'. 'In Burkina Faso, we love Europe—but in the end, we're just left with cheap copies', Kéré explained [37]. This mindset reflects a broader issue where traditional knowledge and materials are dismissed due to colonial influences that have embedded a skewed perception of value in the minds of the Global South.

Figure 5.4 Gando Primary School, Burkins Faso.

Reproduced from Helge Fahrnberger via Wikimedia Commons (2005), https://commons.wikimedia.org/wiki/File:Gando-School-Burkina-Faso.JPG under a Creative Commons Attribution-ShareAlike 3.0 (CC BY_SA 3.0).

This bias extends beyond architecture. Traditional and indigenous knowledge across fields such as agriculture, medicine, and construction has often been mischaracterized as primitive or simplistic. Yet, empirical studies across disciplines are increasingly affirming the sophistication and relevance of this knowledge. Indigenous practices in agriculture and medicine offer insights honed over millennia, optimized for local conditions and sustainability [38, 39]. A good example is one of the 2015 Nobel Prize in Medicine winners, a Chinese scientist named Tu Youyou who discovered a cure for chloroquine-resistant malaria and turned to traditional Chinese medical literature for clues. There, she and her team found a reference to sweet wormwood to treat intermittent fevers [40].

Despite increasing recognition of its value, the field of architectural theory and history continues to underappreciate vernacular architecture, which was often dismissed during the colonial era as merely 'non-expert' construction [41]. Vernacular architecture embodies the traditional and regional design language developed by local communities using indigenous materials and techniques. Yet, it remains marginalized in discussions on sustainable building practices, often pejoratively labelled as 'poor man's knowledge' [42]. The label critically highlights the flawed practice of assessing the merit of knowledge based solely on socioeconomic status, rather than on the actual substance and applicability of the wisdom offered. This oversight is particularly ironic given the proven benefits of vernacular methods, which provide low-cost, low-tech, and highly sustainable solutions exquisitely adapted to local environmental and cultural contexts [43]. The challenge ahead involves a profound re-evaluation of these ingrained biases. It calls for a redefinition of valuable knowledge in the quest for truly sustainable development—a development that respects and fully integrates local wisdom. In Chapter 8, we will delve into how Global South knowledge can help us to find innovative and alternative ways to reduce embodied carbon emissions.

5.4 Green growth: the need for revaluation

The undervaluation of local knowledge often coincides with an overvaluation of so-called 'green technology' or green growth. This bias towards high-tech solutions, predominantly propagated by the Global North, extends beyond mere research and policy implications; it underpins broader power dynamics and establishes relations of exploitative dominance between the Global North and the Global South. This disproportionate influence of the Global North in shaping sustainability agendas not only influences the environmental initiatives but also underlines a pressing need to critically reassess and rectify the biases embedded within current sustainability practices and decision-making processes. Such a critical examination is essential to ensure that sustainability becomes a universally beneficial pursuit rather than a tool for furthering geopolitical and economic hegemonies [44].

5.4.1 Rethinking sustainability: bridging the Global North–South divide

The Global North dominates sustainability research and decision-making, which is evident from the distribution of scientific contributions and topics largely centred on Northern issues [45]. This focus extends to the discourse around sustainable cities, often shaped by the experiences of Northern urban centres. The Sustainable Development Goals (SDGs), launched in 2015 to foster global sustainability, have unfortunately deepened the North–South divide, reflecting a perspective that often fails to accommodate the diverse realities across regions [46, 47].

For example, widely promoted actions like tree planting, biking, driving electric cars, or composting are touted as universal remedies. These measures emphasize market and corporate involvement, pushing for investment in renewable energy and green infrastructure [3]. However, this approach frequently overlooks broader environmental and social repercussions. A prominent example is the core of many renewable technologies: battery production from cobalt mining, which is crucial for electric vehicles and other smart devices. The advocacy for electric vehicles often neglects the severe environmental impacts and harsh conditions associated with cobalt mining, particularly in the Democratic Republic of Congo (DRC), the world's largest cobalt supplier. It is estimated that up to 70% of cobalt production in the DRC comes from artisanal miners [48], yet 73% of the Congolese population lived below the international poverty line of $2.15 a day in 2023 [49].

Cobalt mining is linked to dangerous exposure to harmful dust and grit, negatively impacting lung health [50]. There is also evidence of increased birth defects in children born to parents who work in cobalt mines, likely due to toxic contamination from the extraction process. Additionally, cobalt mining leads to soil degradation, water pollution, and contamination of agriculture. Siddharth Kara's 2023 book, *Cobalt Red: How the Blood of the Congo Powers Our Lives*, reveals the immense toll on the people and environment of the DRC. In an interview with America's National Public Radio, Kara highlighted these severe consequences, underscoring the need to address the hidden costs of the push for green technologies:

> 'We shouldn't be transitioning to the use of electric vehicles at the cost of the people and environment of one of the most downtrodden and impoverished corners of the world,' he says. 'The bottom of the supply chain, where almost all the world's cobalt is coming from, is a horror show.' [51]

As such, questions have arisen as to whether the transition to electric vehicles and cobalt-based batteries is yet another example of environmental problem-shifting and what a renewable energy transition without cobalt could look like. Another great example of such a Global North solution leading to unintended consequences is solar pumps. The adoption of solar pumps in regions like India and Africa showcases the benefits and challenges of such technology [52]. These devices, which operate without cost throughout the day, aid in extending arable land and reducing dependency on inconsistent rainfall [53]. Yet, they also risk depleting already scarce aquifers. The World Bank has highlighted the dangers of the unregulated use of solar pumps, particularly in India, where they exacerbate high groundwater consumption. While intended to replace inefficient diesel pumps and reduce emissions, these pumps often supplement existing ones, leading to intensified farming practices that significantly increase carbon footprint through additional use of fertilizers and heavy machinery [52]. What is more significant is that a solar pump solves the symptoms, but not the causes to problems. It solves the problem of the lack of stable energy supply for pumps and reduces dependence on fossil fuel; however, it did not solve unsustainable farming practices that are water intensive and unsuitable for locations that are hot and dry. A real innovative approach is to extract knowledge from indigenous practices in those regions that have been honed from thousands of years of wisdom, such as no-tillage farming in Iran, pitcher irrigation (clay pot irrigation) in India and Africa [54], bamboo dripping system in India and China, and waffle gardens used for over a thousand years by Zuni people, a Native America Pueblo tribe. Waffle gardens are a traditional water management technique for desert farming that is still in practice today. These gardens feature clay or adobe walls approximately 30–50 cm high that encircle the cultivation plots, often enhanced with a gravel mulch layer to retain soil moisture and minimize evaporation, as illustrated in Figure 5.5. These examples highlight how communities can effectively use local materials and traditional knowledge to create sustainable living solutions without depending on advanced technology.

Over the past five years, the author has attended numerous conferences and workshops where the primary focus was on emission cuts, recycling, and the circular economy. These discussions were held in comfortable, air-conditioned rooms, far removed from the realities they aimed to address, with critical issues like poverty, environmental justice, and governance rarely discussed. This reflects a conventional view of sustainability where solutions are often top-down and do not fully engage with the communities they affect nor respect the local knowledge.

Addressing these challenges requires a profound engagement with indigenous knowledge and practices refined over centuries. Leaders and educators from the Global South promote alternative paths that honour local knowledge and realities. A study in Iran illustrates how local farmers have adapted traditional methods to manage drought conditions, employing strategies like no-tillage farming, diversifying livestock, and indigenous water conservation techniques such as mulching and altering cropping patterns [55]. Similar sustainable and culturally attuned practices are observed in Africa, demonstrating their effectiveness [56]. These analyses emphasize the need to move beyond

Figure 5.5 (a) Waffle garden. Reproduced from National Archives at College Park via Wikimedia Commons, https://commons.wikimedia.org/wiki/File:Vegetable_gardens_surrounding_the_Indian_Pueblo_of_Zuni,_New_Mexico_1873_-_NARA_-_519765.tif; (b) modern day waffle garden.

Reproduced with permission from Ryan, W., 'Sustainable Gardening and Service Learning at the University of Redlands', *Journal of Green Building* (2015) 10(2): 64–78, doi:10.3992/jgb.10.2.64.

a Global North model of sustainability to embrace a nuanced approach that respects and integrates Global South local insights and practices [3].

5.4.2 The tale of green growth

Green growth emerged as a promising concept, advocating that technological innovation could enable an economy to grow while simultaneously reducing environmental impacts. However, the feasibility and definitions of green growth, including the idea of decoupling economic growth from environmental degradation, remain subjects of ongoing debate and have yet to be empirically validated on a global scale.

Around the time of the Rio+20 Conferences in 2011 and 2012, three major institutional proponents of green growth theory—the Organization for Economic Cooperation and Development (OECD), the United Nations Environment Program (UNEP), and the World Bank—each published flagship reports to advance global sustainability. The OECD's 'Towards Green Growth' advocated for economic growth that ensures natural assets continue to provide essential resources and services [57]. During the same period, UNEP released 'Toward a Green Economy: Pathways to Sustainable Development and Poverty Eradication', offering strategies for environmentally friendly economic practices [58]. The World Bank's 'Inclusive Green Growth: The Pathway to Sustainable Development' described green growth as economic progress that is resource-efficient, minimizes environmental impacts, and is resilient to natural hazards [59]. During the Rio+20 Conference, these institutions, alongside the Global Green Growth Institute, launched the Green Growth Knowledge Platform to further global green growth strategies.

The consensus among these leading global institutions underscores a preference for policies that assert the possibility of absolute decoupling, as articulated by UNEP. This notion—that economic expansion can completely detach from increased resource use and environmental impact—has shaped policies primarily in the Global North and has been vigorously promoted in the Global South. However, the metrics used to quantify decoupling raise critical questions about the integrity of green growth claims. 'Domestic material consumption' (DMC) is the standard metric for assessing a nation's resource utilization, accounting for the total mass of raw materials extracted domestically plus all physical imports minus exports. Though not a direct measure of ecological pressure, DMC is a commonly accepted proxy, providing a foundation for policy discussions due to its empirical reliability [60]. Governments often use the GDP-to-DMC ratio to gauge 'resource efficiency', with a higher ratio suggesting an economy is becoming more efficient relative to its resource use [61]. As early as 2017, the Green Growth Indicators report by the OECD highlighted some nations achieving relative decoupling, where GDP rises faster than DMC, indicating an improvement in material productivity, particularly within European OECD countries. The report also suggested that these countries have achieved absolute decoupling, enhancing GDP while simultaneously reducing DMC. Notably, non-energy material consumption within the OECD decreased from 12 tonnes per capita in 2000 to 10 tonnes in 2015, suggesting an optimistic trend of reaching 'peak stuff' [62, 63].

However, DMC as an indicator has its fundamental flaws. For most, it does not account for consumption-based emissions and impact; it is often described as embodied carbon and energy. In this way, developed countries in the Global North have largely outsourced their production to less affluent Global South countries, effectively offloading this aspect of their material consumption.

This results in a stark disparity between production-based and consumption-based carbon emissions. When considering the total resource impact of consumption—the 'material footprint'—the scenario changes dramatically [64]. As illustrated in Figure 5.6, research by Wiedmann and colleagues shows that for nations like the USA, UK, Japan, the OECD, and EU-27 between 1990 and 2008, their material footprint (resource consumption) increased by over 50% and much faster than their GDP output. That indicates no true decoupling—and in most cases, a recoupling. This analysis prompts a critical re-evaluation of the feasibility of green growth, suggesting that the concept of a green economy might be based on misleading metrics. A truly comprehensive measure of economic growth must encompass both production and consumption emissions to fully understand and mitigate the global impacts of material consumption.

When developed countries in the Global North portray a rosy 'green' picture of their growth by partially counting their resource consumption and environmental impact using production-based metrics, the empirical data show a different but more complete picture. Data indicate a dramatic rise in global resource consumption, with the extraction and use of materials, including fossil fuels, increasing twelvefold from 1900 to 2015, reaching 89 billion tons annually [**60**]. This period also saw emissions and waste flows surge by 58 billion tons per year [**65**]. Along with this significant increase, global consumption surged by 93.4% between 1980 and 2009, reaching 67.6 billion tons [**66**].

The feasibility of completely decoupling economic growth from resource use and carbon emissions on a global scale remains unsupported by substantial empirical evidence. While relative decoupling—reducing resource use

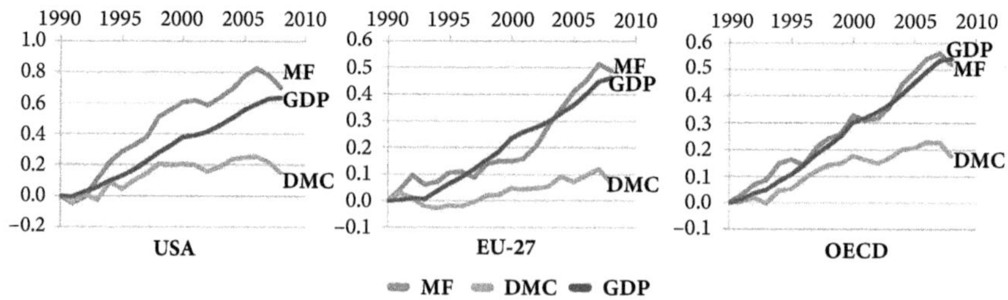

Figure 5.6 Material use trends for USA, EU-27, and OECD, 1990–2008. MF: material footprint; DMC: domestic material consumption; GP: gross domestic product.

Reproduced with permission from Wiedmann, Thomas O. et al. 'The Material Footprint of Nations' (2013) Washington DC, USA: PNAS, doi: https://doi.org/10.1073/pnas.1220362110

relative to GDP growth in certain regions, such as Global North—might be achievable, absolute decoupling, which entails reducing total resource use and emissions while simultaneously growing the economy, appears highly improbable at the rates necessary to prevent catastrophic climate impacts [61].

The fundamental flaw in the green economy or green growth theory is its premise that human activities and economic expansion can be decoupled from the physical limits of the natural ecosystem. While initiatives like solar pumps are steps towards sustainability, they also reveal the shortcomings of relying solely on technological solutions. This raises the question: is there an alternative model that addresses these fundamental issues more directly? Section 5.4.3 tries to provide answers.

5.4.3 Alternative economic models: Herman Daly's steady-state economics

We don't need to search far for a viable pathway; instead, we should reconsider a concept long criticized and dismissed by neoclassical economists: the steady-state economy. Ecological economist Herman Daly, a former senior environmental economist at the World Bank, has been a pivotal figure in challenging the conventional economic growth model through his advocacy for a steady-state economy since 1977. Daly was one of the first prominent voices to counter the overly optimistic claims from neoclassical theory about the potential for decoupling GDP from environmental impact [67]. He has consistently emphasized the inseparability of GDP from its material dimension, arguing that continuous economic expansion and the increased throughput of materials and energy are physically unsustainable on our finite planet [68, 69]. He points out:

> The entire evolution of the biosphere has occurred around a fixed point—the constant solar-energy budget. Modern man is the only species to have broken the solar-income budget constraint, and this has thrown him out of equilibrium with the rest of the biosphere. Natural cycles have become overloaded, and new materials have been produced for which no natural cycles exist. Not only is geological capital being depleted, but the basic life-support services of nature are impaired in their functioning by too large a throughput from the human sector. [68]

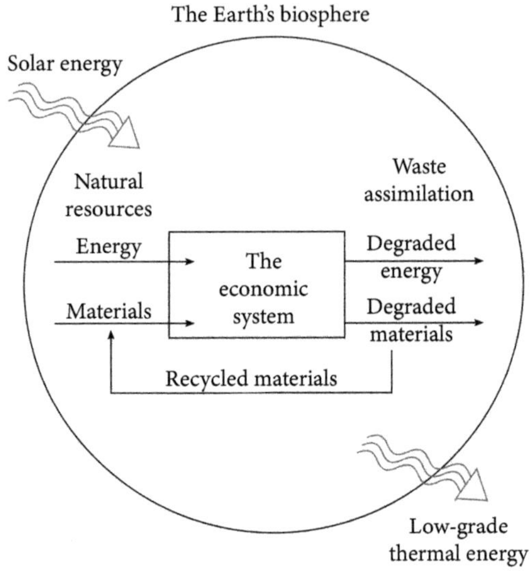

Figure 5.7 Herman Daly's ecological economy diagram.

Reproduced from Gaeanautes via Wikimedia Commons (2018), https://commons.wikimedia.org/wiki/File:Diagram_of_natural_resource_flows.jpg under a Creative Commons CC0 1.0 Universal Public Domain Dedication (CC0 1.0)

Daly contends that the global economy has already exceeded the sustainable capacity of the biosphere's renewable and nonrenewable resources, as illustrated in Figure 5.7. In response, Daly advocates for the immediate establishment of a system of permanent government restrictions to transition to a steady-state economy. Unlike neoclassical economists, who believed that a final stationary state would naturally occur as the rate of profit fell and capital accumulation ceased, Daly proposes a proactive approach. His framework for sustainable development aims to align economic activity with the planet's ecological limits, challenging neoclassical economic paradigms and emphasizing the need to integrate growth limitations.

To implement a steady-state economy with growth limits, Daly suggests creating the three governmental institutions listed in Box 5.2.

Box 5.2: Three governmental institutions for a steady-state economy

- The first would address economic inequality by setting minimum and maximum income and wealth limits and redistributing resources accordingly.

- The second would stabilize the population through the issuance of transferable reproduction licenses at a level that maintains replacement fertility.
- The third would regulate the amount of capital by selling depletion quotas that limit the flow of resources through the economy, thereby minimizing the necessary throughput to maintain any given level of capital. Unlike taxes that adjust prices, quotas directly control the volume of resources used.

Daly envisions a steady-state economy where population and consumption levels are stabilized within the regenerative and assimilative capacities of the ecosystem. This approach emphasizes maintaining an optimal, constant stock of wealth and population for long-term sustainability [69]. Furthermore, Daly's ideas have influenced the concept of carbon trading, a practice now established in regions such as the EU, China, parts of North America, New Zealand, and South Korea. This system, reinforced by the Paris Agreement rulebook, allows for the development of carbon pricing mechanisms and facilitates integration into the emerging global carbon markets, marking a significant step towards globally coordinated environmental governance. Chapter 7 gives a detailed discussion on the carbon trade.

The importance of Daly's framework lies in its recognition of ecological limits and the need to harmonize economic activity with the planet's capacity. Unlike the current green economy approach, which often focuses on technological solutions and relative decoupling, Daly's steady-state economy addresses the root causes of environmental degradation by limiting material and energy throughput. This approach ensures that economic development does not come at the expense of ecological stability. By prioritizing sustainability, equity, and long-term ecological health, Daly's framework offers a more realistic and ethical pathway to achieving a truly sustainable future. But why wasn't Daly's proposed steady-state economy pathway adopted? Daly's proposal not only challenged the neoclassical economy notion that perpetual economic growth is compatible with environmental conservation, but he was also criticized by those sympathetic with his challenge against the growth paradigm. One major critique centres on his reliance on a neoclassical economic framework. Daly's approach attempts to integrate neoclassical concepts like 'uneconomic growth' and market efficiency for resource allocation. Critics argue that this reliance is insufficient for addressing the dynamics of a capitalist economy, which inherently pursues relentless growth. This creates internal inconsistencies within steady-state economics, as Daly adopts neoclassical assumptions that many believe are fundamentally at odds with ecological economic goals.

Another critique focuses on Daly's lack of comprehensive analysis regarding the growth dynamics inherent in capitalism. Critics highlight that simply identifying an optimal economy size, where marginal costs equal marginal benefits, overlooks deeper systemic issues. They contend that Daly's framework does not adequately address how institutions entrenched in capitalist economies drive continuous growth, making it difficult to halt growth without significant and contentious economic transformations [70].

Additionally, Daly's perspectives on population and immigration have been contentious. His proposals, such as tradeable birth licenses, though well-intentioned, have faced criticism for their practicality and ethical implications. His stance on immigration, aimed at maintaining stable population levels, has also sparked debate, particularly concerning its implications for social justice and equity. Critics argue that these aspects of his work reveal a need for a more nuanced approach that considers the complex socioeconomic realities influencing population and migration dynamics [67].

Overall, while Daly's foundational work in ecological economics is invaluable, these critiques underscore the need for further refinement and adaptation of his theories. Addressing these criticisms involves moving beyond neoclassical economic foundations and developing more robust theories that incorporate realistic assumptions about economic growth, power dynamics, and social equity. This evolution is crucial for creating effective policies and triggering transformative changes towards sustainability.

5.5 Conclusion

Chapter 5 delves into the complex interplay between economic theories, environmental sustainability, and geopolitical influences. It begins by critically analysing the Global North bias, illustrating how this perspective skews global policies and practices, particularly in sustainability and environmental justice. The chapter highlights how the priorities and methodologies of the Global North dominate sustainability dialogues, often sidelining the unique circumstances and needs of the Global South. This bias is evident in sustainability metrics and strategies that fail to fully address the realities faced by less developed nations. The discussion on green growth theories and the promotion of these ideals by major global institutions like the OECD, UNEP, and the World Bank reveals a tendency to prioritize economic growth over ecological balance.

The focus then shifts to alternative economic models, with Herman Daly's steady-state economy theory emerging as a critical counterpoint. Daly advocates for restructuring economic systems to align with the planet's ecological limits, proposing government interventions to stabilize economic activity within these boundaries. The chapter concludes by calling for a fundamental shift in how economic growth and sustainability are perceived and pursued. It emphasizes the need to adopt models that truly integrate ecological and social equity considerations, essential for moving towards a sustainable future. This re-evaluation is presented as crucial for ensuring that economic activities do not exceed the planet's ecological thresholds.

As we transition to Chapter 6, the focus shifts to another significant influence: high-tech bias. This bias favours technologically advanced solutions in environmental strategies, often overlooking simpler, potentially more effective methods. The exploration of high-tech bias will reveal how it mirrors and extends the complexities discussed in green growth and steady-state economics, potentially reinforcing existing inequities and hindering true sustainability.

References

1. Share The World's Resources. The Brandt Report: a summary [Internet]. [Accessed 29 Apr 2024]. Available from https://sharing.org/information-centre/articles/brandt-report
2. Nagendra H, Bai X, Brondizio ES, Lwasa S. The urban south and the predicament of global sustainability. Nature Sustainability. 2018; 1(7): 341–9.
3. Nagendra H. The global south is rich in sustainability lessons that students deserve to hear. Nature. 2018 May; 557(7706): 485–8.
4. Singh P, Bajwa A. Who is responsible for the climate crisis? A perspective from the Global South. International Critical Thought. 2023 Jul 3; 13(3): 429–44.
5. Chen Y. How has ecological imperialism persisted? A Marxian critique of the Western climate consensus. American J Econ Sociol. 2022 May; 81(3): 473–501.
6. Yin I, Yep E. Feature: IEA's net-zero roadmap is unrealistic, lacks localized approach - Chinese experts [Internet]. 2021 Jun [accessed 29 Apr 2024]. Available from https://www.spglobal.com/commodityinsights/en/market-insights/latest-news/natural-gas/061121-feature-ieas-net-zero-roadmap-is-unrealistic-lacks-localized-approach-chinese-experts?utm_campaign=carbon%20brief%20daily%20briefing&utm_content=20210614&utm_medium=email&utm_source=revue%20daily

7. You X. Analysis: What does China's coal push mean for its climate goals? [Internet]. Carbon Brief. 2024 [accessed 2024 Jun 24]. Available from https://www.carbonbrief.org/analysis-what-does-chinas-coal-push-mean-for-its-climate-goals/
8. bp. bp statistical review of world energy 2021 [Internet]. London; 2021 [accessed 24 Jun 2024]. Available from https://www.bp.com/content/dam/bp/business-sites/en/global/corporate/pdfs/energy-economics/statistical-review/bp-stats-review-2021-full-report.pdf
9. International Energy Agency. Renewable electricity growth is accelerating faster than ever worldwide, supporting the emergence of the new global energy economy [Internet]. 2021 Jan [accessed 24 Jun 2024]. Available from https://www.iea.org/news/renewable-electricity-growth-is-accelerating-faster-than-ever-worldwide-supporting-the-emergence-of-the-new-global-energy-economy
10. U.S. Energy Information Administration. Solar and wind to lead growth of U.S. power generation for the next two [Internet]. 2024 [Accessed 1 Aug 2024]. Available from https://www.eia.gov/todayinenergy/detail.php?id=61242
11. Dale G, Mathai MV, de Oliveira JAP. Green growth: ideology, political economy and the alternatives. Bloomsbury Publishing; 2016.
12. Seager J. Death by degrees: taking a feminist hard look at the 2 climate policy. Kvinder, Køn & Forskning. 2009; (3–4).
13. Parmesan C, Morecroft MD, Trisurat Y. Climate change 2022: Impacts, adaptation and vulnerability. 2022; IPCC Sixth Assessment Report.
14. Loconto A, Fouilleux E. Politics of private regulation: ISEAL and the shaping of transnational sustainability governance. Regulation & Governance. 2014; 8(2): 166–85.
15. Newell P. Power shift: The global political economy of energy transitions. Cambridge University Press; 2021.
16. Pojani D, Stead D. Policy design for sustainable urban transport in the global south. Policy Design and Practice. 2018; 1(2): 90–102.
17. Lewis M. Cattle and conservation at Bharatpur: a case study in science and advocacy. Conservation and Society. 2003; 1(1): 1–21.
18. Jenkins K. Setting energy justice apart from the crowd: Lessons from environmental and climate justice. Energy Research & Social Science. 2018 May; 39: 117–21.
19. Mohai P, Pellow D, Roberts JT. Environmental justice. Annual Review of Environment and Resources. 2009; 34: 405–30.
20. Knoble C, Yu D. Environmental justice: an evolving concept in a dynamic era. Sustainable Development. 2023 Aug; 31(4): 2091–108.
21. Jackson AV. Environmental justice: what's racism got to do with it? Intl J Gynecology & Obste. 2023 Feb; 160(2): 372–3.
22. Allen BL. Environmental justice and expert knowledge in the wake of a disaster. Soc Stud Sci. 2007 Feb; 37(1): 103–10.
23. Zhou X, Guo Q, Wang Y, Wang G. Trade and embodied CO2 emissions: analysis from a global input–output perspective. IJERPH. 2022 Nov 7; 19(21): 14605.
24. Hinojal IB, Aurrekoetxea RL. Ecological debt: an integrating concept for socio-environmental change. In: Redclift MR, Woodgate G, editors. The international handbook of environmental sociology, 2nd edn [Internet]. Edward Elgar Publishing; 2010 [accessed 2024 Apr 30]. Available from https://china.elgaronline.com/view/edcoll/9781848440883/9781848440883.00019.xml

25. Roberts JT, Parks BC. A climate of injustice: global inequality, north-south politics, and climate change. Cambridge, MA: MIT Press; 2007.
26. Walker G, Bulkeley H. Geographies of environmental justice. Geoforum. 2006 Sep; 37(5): 655–9.
27. Matthews HD. Quantifying historical carbon and climate debts among nations. Nature Clim Change. 2016 Jan; 6(1): 60–4.
28. Baba M. Carbon-negative Bhutan offers lessons on fighting climate change [Internet]. NIKKEI Asia. 2022 [Accessed 1 May 2024]. Available from https://asia.nikkei.com/Spotlight/Environment/Climate-Change/Carbon-negative-Bhutan-offers-lessons-on-fighting-climate-change
29. Fanning AL, Hickel J. Compensation for atmospheric appropriation. Nat Sustain. 2023 Jun 5; 6(9): 1077–86.
30. Helm D. Legacy: how to build the sustainable economy [Internet]. 1st edn. Cambridge, UK: Cambridge University Press; 2023 [accessed 6 May 2024]. Available from https://www.cambridge.org/core/product/identifier/9781009449212/type/book
31. U.S Environmental Protection Agency. Global greenhouse gas overview [Internet]. [Accessed 29 Apr 2024]. Available from https://www.epa.gov/ghgemissions/global-greenhouse-gas-overview#:~:text=Emissions%20by%20Country,-Source%3A%20Data%20from&text=In%202020%2C%20the%20top%20ten, change%2C%20industry%2C%20and%20waste.
32. Bouckaert S, Pales AF, McGlade C, Remme U, Wanner B, Varro L, et al. Net zero by 2050: a roadmap for the global energy sector. 2021. Available at https://www.iea.org/reports/net-zero-by-2050.
33. Carbon Brief. Revealed: how colonial rule radically shifts historical responsibility for climate change [Internet]. 2023 [Accessed 6 May 2024]. Available from https://www.carbonbrief.org/revealed-how-colonial-rule-radically-shifts-historical-responsibility-for-climate-change/
34. Schlosberg D, Collins LB. From environmental to climate justice: climate change and the discourse of environmental justice. WIREs Climate Change. 2014 May; 5(3): 359–74.
35. Hickel J. Quantifying national responsibility for climate breakdown: an equality-based attribution approach for carbon dioxide emissions in excess of the planetary boundary. Lancet Planetary Health. 2020; 4(9): e399–404.
36. Holcim Foundation. Prelude to 'Re-materializing Construction' in Cairo [Internet]. 2019 [Accessed 1 May 2024]. Available from https://www.holcimfoundation.org/media/news/foundation/prelude-to-re-materializing-construction-in-cairo
37. – Holcim Forum. Inspired by tradition, adapted to local environment: Francis Kéré on 'Re-materializing Construction' [Internet]. 2019. Available from https://www.holcimfoundation.org/media/news/forum/inspired-by-tradition-adapted-to-local-environment
38. DeWalt B. Using indigenous knowledge to improve agriculture and natural resource management. Human organization. 1994; 53(2): 123–31.
39. Rahal A, Deb R, Latheef SK, Tiwari R, Verma A, Kumar A, et al. Immunomodulatory and therapeutic potentials of herbal, traditional/indigenous and ethnoveterinary medicines. Pakistan journal of biological sciences: PJBS. 2012; 15(16): 754–74.
40. World Health Organization. Traditional medicine has a long history of contributing to conventional medicine and continues to hold promise [Internet].

08/23 [Accessed 6 May 2024]. Available from https://www.who.int/news-room/feature-stories/detail/traditional-medicine-has-a-long-history-of-contributing-to-conventional-medicine-and-continues-to-hold-promise
41. Oliver P. Built to meet needs: cultural issues in vernacular architecture [Internet]. London: Routledge; 2007 [Accessed 23 Nov 2022]. Available from https://www.taylorfrancis.com/books/9781136424052
42. Brown R, Maudlin D. Concepts of vernacular architecture. The SAGE Handbook of Architectural Theory. London: Sage; 2012. pp. 340–55.
43. Aktürk G, Fluck H. Vernacular heritage as a response to climate: lessons for future climate resilience from Rize, Turkey. Land. 2022 Feb 11; 11(2): 276.
44. Korteling JE, Paradies GL, Sassen-van Meer JP. Cognitive bias and how to improve sustainable decision making. Frontiers in Psychology. 2023; 14: 1129835.
45. Kanmodi KK, Amzat J, Nnyanzi LA. The state of global research on poverty after SDG declaration. Public Health Challenges. 2023; 2(3): e110.
46. Gusmão Caiado RG, Leal Filho W, Quelhas OLG, Luiz De Mattos Nascimento D, Ávila LV. A literature-based review on potentials and constraints in the implementation of the sustainable development goals. Journal of Cleaner Production. 2018 Oct; 198: 1276–88.
47. Kundu D, Sharma P, Gupta D, Mueller A, Gareis P. Measuring sustainable urban development: the case of India, Europe and Germany. Environment and Urbanization ASIA. 2023 Sep; 14(2): 269–84.
48. Cobalt Institute. Growing demand for cobalt sourcing—driven by its crucial role in enabling electric mobility and the green economy—puts into the spotlight the way this critical mineral is sourced. [Internet]. [Accessed 4 Aug 2024]. Available from https://www.cobaltinstitute.org/cobalt-sourcing-responsability/#:~:text=Approximately%2070%25%20of%20the%20world's,known%20as%20the%20%E2%80%9CCopperbelt%E2%80%9D.
49. World Bank Group. The World Bank in DRC [Internet]. [Accessed 4 Aug 2024]. Available from https://www.worldbank.org/en/country/drc/overview
50. Venkat K. Cobalt mining in Congo: environmental health, safety, & human rights [Internet]. The Climate Club. [Accessed 4 Aug 2024]. Available from https://www.theclimateclub.co/environmental-healthblogs/347zlomifb9x9jsyogvi4kymlynxfb
51. Kara S. How 'modern-day slavery' in the Congo powers the rechargeable battery economy [Internet]. 2023 [Accessed 4 Aug 2024]. Available from https://www.npr.org/sections/goatsandsoda/2023/02/01/1152893248/red-cobalt-congo-drc-mining-siddharth-kara
52. WIRED. Solar-powered farming is quickly depleting the world's groundwater supply [Internet]. 2024 [Accessed 2 May 2024]. Available from https://www.wired.com/story/solar-energy-farming-depleting-worlds-groundwater-india/
53. Balasubramanya S, Garrick D, Brozović N, Ringler C, Zaveri E, Rodella AS, et al. Risks from solar-powered groundwater irrigation. Science. 2024 Jan 19; 383(6680): 256–8.
54. Adhikary R, Pal A. Clay pot irrigation—a review study. APRJ. 2020 Jun 17; 37–42.
55. Aliabadi V, Ataei P, Gholamrezai S. Farmers' Strategies for drought adaptation based on the indigenous knowledge system: the case of Iran. Weather, Climate, and Society. 2022 Apr; 14(2): 561–8.

56. Muyambo F, Bahta YT, Jordaan AJ. The role of indigenous knowledge in drought risk reduction: A case of communal farmers in South Africa. Jàmbá [Internet]. 2017 Jul 27 [accessed 2 May 2024]; 9(1). Available from https://jamba.org.za/index.php/jamba/article/view/420
57. OECD. Towards green growth: a summary for policy makers [Internet]. 2011 May [accessed 3 May 2024]. Available from https://www.oecd.org/greengrowth/48012345.pdf
58. UNEP. Towards a green economy: pathways to sustainable development and poverty eradication [Internet]. 2011 [Accessed 3 May 2024]. Available from https://wedocs.unep.org/handle/20.500.11822/32245
59. The World Bank. Inclusive green growth: the pathway to sustainable development: main report [Internet]. [Accessed 3 May 2024]. Available from https://documents.worldbank.org/en/publication/documents-reports/documentdetail/368361468313515918/main-report
60. Krausmann F, Gingrich S, Eisenmenger N, Erb KH, Haberl H, Fischer-Kowalski M. Growth in global materials use, GDP and population during the 20th century. Ecological Economics. 2009; 68(10): 2696–705.
61. Hickel J, Kallis G. Is green growth possible? New Political Economy. 2020 Jun 6; 25(4): 469–86.
62. Goodall C. Peak stuff: did the UK reach a maximum use of material resources in the early part of the last decade. A research paper for Carbon Commentary. 2011; 13.
63. Pearce F. Peak planet: are we starting to consume less? New scientist. 2012; 214(2869): 38–43.
64. Wiedmann TO, Schandl H, Lenzen M, Moran D, Suh S, West J, et al. The material footprint of nations. Proceedings of the National Academy of Sciences. 2013; 112(20): 6271–6.
65. Kovanda J. Economy-wide material system analysis: mapping material flows through the economy. J of Industrial Ecology. 2021 Oct; 25(5): 1121–35.
66. Giljum S, Dittrich M, Lieber M, Lutter S. Global patterns of material flows and their socio-economic and environmental implications: a MFA study on all countries worldwide from 1980 to 2009. Resources. 2014; 2(1): 319–39.
67. Hayden A. Reflections on Herman Daly's economics for a full world: his life and ideas. Sustainability: Science, Practice and Policy. 2022 Dec 9; 18(1): 630–5.
68. Daly HE. Steady-state economics versus growthmania: a critique of the orthodox conceptions of growth, wants, scarcity, and efficiency. Policy Sciences. 1974; 149–67.
69. Daly HE. Ecological economics and sustainable development: selected essays of Herman Daly. Cheltenham, UK; Northampton, MA: Edward Elgar; 2007. (Advances in ecological economics).
70. Pirgmaier E. The neoclassical Trojan Horse of steady-state economics. Ecological Economics. 2017 Mar; 133: 52–61.

Chapter 6
High-tech bias

Implications for sustainability

6.1 Define high-tech bias

Technology bias can be defined as automatically activated perceptions of emerging technology [1]. The implicit biases draw from general beliefs about technology, especially emerging, advanced, and high technology, and they go on to influence our perceptions of everything from smartphones to flight instruments used to pilot an aircraft [2]. The common perception is that new technology is mysterious and a 'wonder', which can lead to the bias that favours new technology over previous or current options. The most recent example is artificial intelligence (AI); as the technology giant Bill Gates said during an interview, '[AI] in every field of endeavour, will be accelerating innovation, whether that's in medicine or helping with tutoring, education', and he has confidence that ' AI is going to make [combat] climate [change] far easier to do' [3]. Is it true? To answer that, we must first understand how we perceive technology.

6.1.1 Unpacking the high-tech concept

Dr. Kinberly Elsbach and Dr. Ileana Stigliani, two experts in organizational management and innovation, have identified three common underlying beliefs about new technology: mysterious, nonhuman and complex [1]. The term 'mysterious' often applies to new technologies perceived as novel and advanced, fostering a belief that novelty equates to superiority and success. Nonhuman refers to the new technologies that are viewed as nonhuman, which influences trust and perceptions of reliability. The complexity of new technologies is believed to make them difficult to understand without expert knowledge, enhancing trust when credible sources or experts endorse technologies [1]. These beliefs form the basis of several implicit biases, such as associating new technology with success, viewing endorsed technologies

Embodied Carbon. Ming Hu, Oxford University Press. © Ming Hu (2025). DOI: 10.1093/9780198944485.003.0006

as more trustworthy, and perceiving technologies as predominantly Global North domains.

Expanding from the technology bias, high-tech bias refers to a phenomenon where there is a predisposition or favouritism towards high-tech industries, often resulting in unequal treatment or opportunities compared to non-high-tech sectors. This bias can manifest in various ways, such as in research funding allocation, policy support, or societal perceptions of innovation and success. For instance, in 2023, the United States announced it would provide over $3 billion across agencies to develop, test, procure, and integrate transformative AI applications across the federal government. Also, a $300 million budget was proposed for mandatory funding to increase agency funding for AI [4]. Early research conducted by Stam and Wennberg (2009) highlights how high tech can moderate the relationship between innovation activities and firm growth, indicating a unique influence of technological dynamism on entrepreneurial opportunities [5]. This suggests that the high-tech environment can significantly impact the growth trajectory of firms engaging in innovation. And the favouritism towards innovation underscores the imbalance in its support across different types of business. This bias can lead to disparities in resources and opportunities for non-high-tech small and medium-sized businesses, affecting their performance and growth potential [6].

Meanwhile, the effectiveness of high technology is debatable. A study used data from 2009 to 2016 covering sixteen provinces in China to explore the relationship between green growth and high-tech industries. The findings suggest that while high-tech industries have shown progress in innovation, their environmental impact is mixed. However, there has been a general reduction in pollutants like sulphur dioxide due to technological advancement; the overall effect remains a concern [7]. Additionally, Dr Feng and colleagues' 2023 study shows that high-tech companies may manipulate their research and development (R&D) investments to qualify for tax benefits, indicating strategic exploitation of the high-tech label for financial gain [8]. Another example is bioenergy and biofuel, which have become buzzwords. Policymakers and technologists advocate replacing petrol, diesel fuel, and natural gas with plant oils and biogas. As early as 2011, 169 international scientists warned the EU against so-called 'bio petrol', and do not support the high-tech solution to produce 'biofuel' to achieve climate neutrality. There is not a sufficient quantity of biomass available to make it a substitute for conventional energy sources, and intensifying its use beyond processing residual materials (bio waste, edible oil, etc.) offers only limited potential before the delivery of ecosystem services becomes endangered [9, 10].

But what are the causes of high-tech bias? People's perception and bias towards high tech over conventional low tech is fuelled by extremely wealthy individuals in the Global North who portray themselves as problem solvers and thought leaders. Quite a few leaders in the high-tech industry advocate for more technological advancement, even 'miracle' technology. Bill Gates was adamant that we must put a lot of money into miracle technology R&D, which fuelled the concept of 'green growth' and 'decoupling' environmental impact from economic growth. In 2014, he argued that the federal government should fund long-term research to $16 billion yearly as if money and technological advancement were the only solution [11]. One of the high-tech Gates promotes is carbon storage and carbon capture technology, which has become a very hot trend. Let's unpack the carbon capture technologies.

People are probably unaware that carbon capture technology is not new; the first carbon capture plant was proposed in 1938, and the first large-scale project to inject CO_2 into the ground was launched in the Sharon Ridge oilfield in Texas (US) in 1972 [12]. Around twenty-four years later, Norway launched the world's first integrated carbon capture and storage project, known as Sleipner, in the North Sea [12]. However, in its fifty years of existence, carbon capture globally has sequestered only about 40 million metric tons of CO_2 each year, which is less than 1% of global greenhouse gas emissions. This modest impact is further marred by numerous project failures [13].

Carbon capture technology encompasses a variety of methods aimed at capturing CO_2 emissions from different sources, such as power plants and industrial processes, to prevent their release into the atmosphere. These technologies typically involve capturing CO_2, compressing it, transporting it, and then either storing it permanently underground or utilizing it for other purposes. The three commonly used technologies are carbon capture and storage (CCS), direct air carbon capture (DACC), and blue hydrogen capture. CCS typically captures CO_2 from flue gases after fossil fuels have been burned, and then compresses, transports, and stores it underground in rocks [14]. DACC uses chemical processes to extract CO_2 directly from the atmosphere, which is less efficient and more expensive than CCS [15], then compresses, transports, and stores CO_2. Blue hydrogen capture focuses on producing hydrogen from natural gas while capturing and storing the resulting carbon emissions, making it a cleaner alternative to traditional hydrogen production methods, though it is not as mature as the previous two technologies [16].

Thought leaders in another camp have voiced different options from developing more and newer technology. Stanford professor Mark Jacobson suggests that carbon capture technologies can cause more harm than good. His study shows that carbon capture has a net reduction of only 10.8% over twenty years and 20% over a hundred years [17], and twenty years is a relevant time to avoid 1.5°C global warming and resulting climate change [18]. This means that carbon capture technologies alone will not be sufficient to avert climate change, and certainly, it is no miracle technology.

The low net capture rates result from uncaptured emissions originating from three main sources: the combustion of natural gas used to power the equipment, upstream emissions associated with fuel extraction and processing, and residual coal combustion emissions from the carbon capture plant. Moreover, all carbon capture plants increase air pollution and total social costs relative to no capture [17]. Another concern Jacobson raised is the potential environmental risks associated with carbon capture and storage. These include the possibility of CO_2 leakage from storage sites, which could undermine the effectiveness of the technology and pose risks to surrounding ecosystems and communities. He also points out that the infrastructure required for carbon capture, such as pipelines and storage facilities, could have negative environmental impacts. Moreover, Jacobson also argues that focusing on carbon capture technologies can divert attention and investment away from renewable energy solutions. He believes that prioritizing carbon capture technology could slow down the transition to clean energy by providing a false sense of security that fossil fuel use can continue without significant consequences. He advocates for policies and investments that support the rapid deployment of wind, solar, and other renewable energy sources. As Jacobson states:

> [W]e don't need 'miracle' green technologies to save the planet... we have the technologies that we need. We have wind, solar, geothermal, hydro, electric cars. We have batteries, heat pumps, energy efficiency. We have 95% of the technologies right now that we need to solve the problem. [19]

6.1.2 Examples of high-tech bias in sustainability

Concerns within the technologist camp about high-tech solutions like carbon capture highlight the uncertainty of whether any new or advanced technology alone can solve complex problems. A prime example of high-tech bias in sustainability is the prevalent preference for active systems over passive ones to enhance building energy efficiency. While advanced technologies like solar panels are often seen as the ultimate green solutions, they should ideally be considered only after implementing fundamental passive energy-saving measures [20, 21]. Passive design strategies—such as natural ventilation, daylighting, and thermal mass, which absorbs, stores, and releases heat—effectively utilize natural resources to minimize carbon emissions by reducing operational energy demands. Historically, these strategies have enabled humans to live comfortably in harsh climates long before industrial technology, emphasizing human-centred and health-conscious design principles [20, 22, 23, 24, 25, 26]. A systematic literature review conducted by the author and her research collaborators identified thirty passive low-cost or no-cost cooling strategies, showing that employing various passive strategies can lead to an average indoor temperature decrease of 2.2°C, a cooling load reduction of 31%, energy savings of 29%, and a thermal comfort hour extension of 23% [21]. These findings demonstrate the effectiveness of no- and low-cost passive cooling strategies, underscoring the importance of prioritizing passive measures before resorting to high-tech solutions.

Today, there is a clear shift towards optimizing the efficiency of active systems, rather than focusing on reducing overall energy demand. While advancements in technology have driven improvements in energy efficiency, U.S. electricity consumption is still expected to grow—albeit slowly—at an average annual rate of about 1% from 2022 to 2050 [27]. The increasing prevalence of air conditioning in American homes underscores this trend: from just 57% in 1980 to nearly 89% in 2020, spurred by technological progress that, ironically, could lead to greater energy use overall [28]. But these numbers prompt important questions: is this technology reaching everyone? Who has been left behind in this drive towards efficiency? Consider the 11% of U.S. households that still lack air conditioning—what are their stories?

6.1.3 Is high-tech for everyone?

Let's continue with the example of air conditioning technology. As climate change exacerbates heat waves, more people living in houses without passive cooling strategies have come to rely heavily on air conditioning, especially in the United States. This over-reliance presents three significant issues. First, it leads to a surge in energy use; in 2020, air conditioning accounted for about 19% of electricity consumption in U.S. homes [29]. Second, it strains power grids, especially during heatwaves, risking power outages when demand exceeds supply. Third, it highlights environmental injustice; climate change intensifies weather extremes, disproportionately affecting marginalized groups who cannot afford air conditioning or opt out of using it due to the high energy burden [30].

Energy burden refers to the percentage of income a household spends on energy bills. Research shows that households near or below the federal poverty line in the U.S. bear a significant energy burden [31]. This burden is particularly pronounced among vulnerable groups such as Black, Hispanic, low-income, elderly, and those living in energy-inefficient buildings [32]. Low-income households face a substantial disparity in energy costs compared to wealthier households, spending more than four times as much on electricity and over three times as much on heating [33]. Efforts to address energy inequality and reduce the energy burden among low-income households are crucial for promoting social equity and improving overall well-being. By moving away from a high-tech reliance mindset and implementing effective passive strategies, designers and policymakers can work towards alleviating the disproportionate impact of energy costs on low-income families.

On the other side of the ocean, European leaders have taken a different approach by restricting air conditioning usage rather than expanding it. For example, Spanish regulations mandate that air conditioning settings in public spaces can't drop below 26.6°C, a policy mirrored in Italy and Greece. Meanwhile, in the United States, the typical summer thermostat setting is around 24.4°C [34]. While some may think this different approach in Europe is due to environmental consciousness and cultural preferences, cost concerns are actually the main driver, just as in other parts of the world. Public opinion in France provides insights into resistance against adopting air conditioning during severe heatwaves. Air conditioning in Europe is subject to a 20% tax, compared to only 5% for new heating systems. Although environmental and health concerns are often cited as reasons for avoiding air conditioning, these do not fully explain the prevalence of energy-intensive heating systems. It's crucial to note that while many public and office buildings in Europe are equipped with air conditioning, the high cost and associated electricity bills are a major deterrent for residential units.

Another high-tech example is the adoption of solar panels. As discussed in Chapter 5, solar energy adoption exacerbates demographic inequalities due to high initial costs and uneven access to incentives and loans. A study conducted by Wang et al. at Stanford University using satellite imagery to assess solar panel installations revealed that installations are markedly less frequent in low-income American neighbourhoods [35]. Similarly, high-technology solutions like heat pumps and other electrification upgrades often see lower adoption rates in communities with higher proportions of racial and ethnic minorities, lower median incomes, and a higher percentage of renters [36]. This disparity is especially pronounced in areas where heat pumps would otherwise provide substantial benefits to households [37].

6.2 High-tech bias in the built environment

High-tech bias in the built environment often emphasizes advanced technological solutions at the expense of simpler, low-tech alternatives that are inherently more sustainable and cost-effective for addressing challenges like energy use and carbon emissions [38]. For example, traditional and vernacular architecture relies on low-cost, environmentally friendly design principles that are less energy-intensive than modern methods [39]. These practices typically utilize locally available materials and construction techniques familiar to local builders, maintaining regional identities through distinctive architectural styles [40, 41], such as windcatchers used in hot and dry climates.

Since the Industrial Revolution, there has been a shift towards 'high-tech' solutions in the built environment, characterized by heavily processed and standardized materials such as steel, concrete, and glass [42]. This shift has contributed to the homogenization of design and construction techniques, leading to a widespread adoption of the International Style [41, 42]. This style, dominant in the Global North countries by the 1970s and continuing today, is often associated with the 'architecture of the modern movement' or the 'architecture of the machine age' [43], symbols of technological advancement with significantly higher embodied carbon than traditional materials like wood and stone. Figure 6.1 shows an example of a typical International Style building—the Equitable Building in Atlanta, Georgia, constructed in 1968. This building exemplifies the ubiquitous nature of this architectural style across the Global North, which is celebrated for embodying the American dream [44] and is synonymous with continuous growth and innovation [42].

The preference for 'innovative' technology over traditional, low-cost solutions can sometimes lead to adverse human impacts. In 2016, Charles Munger, vice-chairman of Warren Buffett's holding company, proposed building a windowless dormitory at the University of California, Santa Barbara. Munger's design, intended to cut costs and encourage social interaction in communal spaces lit by 'real sunlight', proposed using artificial windows (backlit glass

Figure 6.1 Equitable building at Atlanta, Georgia, U.S. Constructed in 1968.
Reproduced from Tom Caiafa via Wikimedia Commons (2009) under a Creative Commons Attribution 2.5 (CC BY 2.5).

panel) that could simulate natural light changes, replacing the benefits of real windows. The proposed building is eleven stories with just two entrances and would house 4,500 students. This plan faced significant backlash for overlooking basic human and environmental needs, leading to criticism from the architectural community and the public. Notably, Dennis McFadden, an architect on the university's design review committee, resigned in protest, declaring the project 'unsupportable from my perspective as an architect, a parent, and a human being' [45]. However, the billionaire investor kept defending the windowless artificial lighting panel as 'actually better' than the real one without any evidence [46]. Eventually, due to public resistance and criticism over the proposed design, the plan was aborted.

Interestingly, this wasn't Munger's first attempt at a 'high-tech fake window' design; he previously funded the construction of the Munger Graduate Residences at the University of Michigan (Figure 6.2). The building, which opened in 2015, received mixed reviews from residents, some of whom expressed discontent with the lack of windows—a sentiment humorously echoed by a

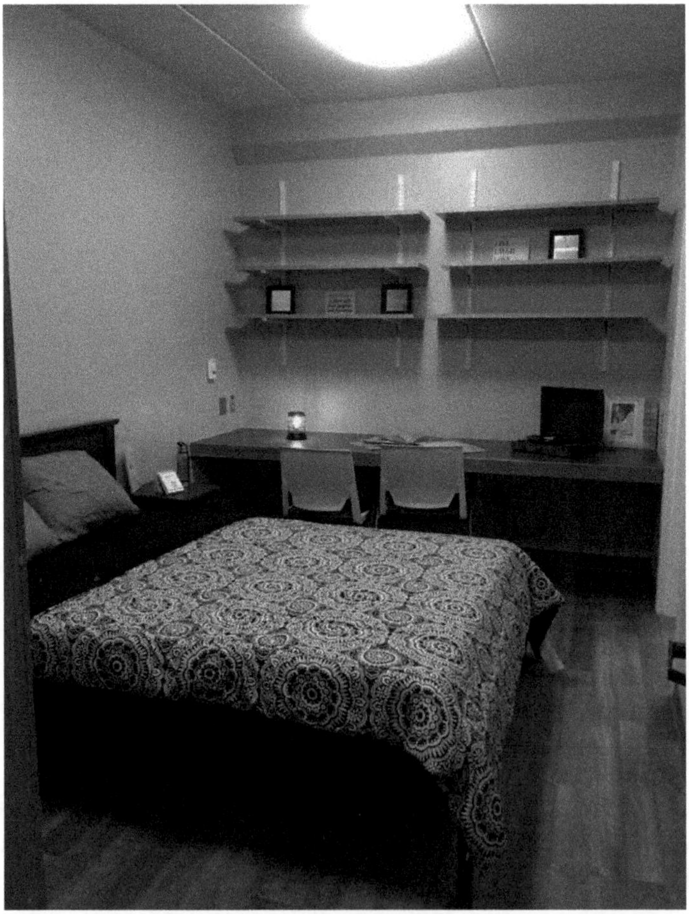

Figure 6.2 Munger Graduate Residences at the University of Michigan.

former student on social media, who questioned whether the absence of windows could genuinely foster friendships. Other students did not shy away from criticizing the windowless design; in many cases, students and former residents have complained that lack of natural light affects their sleep and leads to seasonal affective disorder and depression [47]. All of those underscore the disconnect between Munger's high-tech, biased architectural philosophy and the preferences of the building's occupants [48].

As we delve deeper into the influence of technology on the built environment, it is essential to examine the multifaceted impacts this bias has on various aspects of building design and construction. The following subsections will explore these influences in greater detail. Firstly, we will discuss the influence of technology on building design, highlighting how technological advancements have transformed architectural practices and the implications of these changes. Secondly, we will explore data-driven decision-making in architecture and construction, examining how integrating big data and analytics reshapes the industry. Finally, we will consider the impact of smart technologies, investigating their role in modern construction and the potential benefits and challenges they bring. Together, these sections will provide a comprehensive understanding of how technology is shaping the future of the built environment.

6.2.1 The influence of technology in building design

The incorporation of advanced technology in building design has significantly transformed architectural practices and construction methods. Architects and designers now routinely employ sophisticated technology to enhance indoor air quality and optimize natural lighting, aiming to create pleasant and healthy environments for occupants. However, this technocentric approach often favours newer or high-cost solutions at the expense of more sustainable or locally adapted methods, potentially sidelining traditional practices better suited to specific environmental and cultural contexts.

The profound impact of modern technological interventions is especially evident in the evolution of apartment building designs, which showcase some of the most visible and significant changes in the built environment. Originally, mid- to high-rise apartment buildings incorporated traditional design principles aimed at providing natural cooling through shade, promoting air circulation, and maximizing daylight and views through operable windows and other design features. Historic high-rise structures in cities like Chicago and New York were shaped by those passive design principles before the widespread adoption of technologies such as air conditioning.

These buildings were designed with high ceilings, operable windows, shallow floor plates, and extensive perimeter exposure to enhance ventilation and natural light. Many featured central or side open courts and light wells, forming shapes like 'C', 'E', or 'H' to ensure every unit had access to daylight and cross-ventilation, as illustrated in Figure 6.3. Additionally, because mechanical ventilation was not yet common, kitchens and bathrooms were located close to the building's exterior wall to include operable windows to expel odours naturally, as illustrated in Figure 6.3.

This sustainable and energy-efficient approach to building design began to fade with the advent of air conditioning, mechanical ventilation, and electric lighting, all of which consume energy. Later architectural plans, depicted in Figure 6.4, typically employed double corridors with back-to-back units to maximize centralized cooling systems' efficiency, making natural ventilation largely obsolete. Kitchens and bathrooms were relocated to the core of the building, devoid of windows, as mechanical ventilation systems handled odour control. Consequently, building depths increased, reducing the penetration of natural light and elevating dependence on artificial lighting solutions. This shift marks a significant departure from earlier, more sustainable practices, reflecting the broader implications of technology-driven design in modern architecture. Those internal changes were reflected on the buildings' exterior, making the building look more homogenous rather than with individual characteristics.

6.2.2 Data-driven decision-making in architecture and construction

Data-driven decision-making has transformed architecture and construction, shifting from intuition to sophisticated data analysis to guide strategic choices [49]. This transition has integrated cutting-edge technologies like big data and analytics into the field, impacting key areas such as predictive analytics, building information modelling (BIM), safety, risk management, and sensor data analysis [50, 51].

Predictive analytics leverages statistical modelling, machine learning, and data mining to interpret historical data and forecast future outcomes of construction projects. This powerful approach enables accurate predictions about project performance, potential risks, and the overall feasibility of project

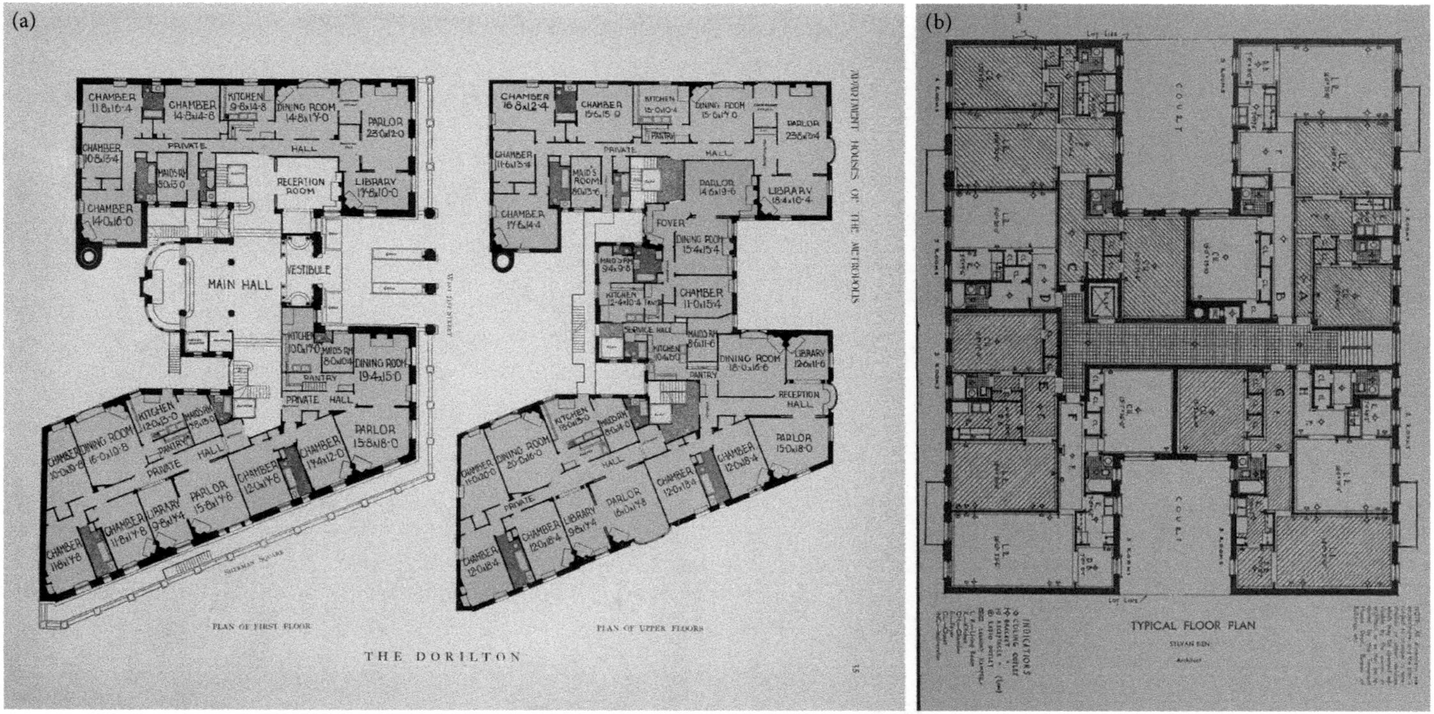

Figure 6.3 Pre-World War II apartment buildings. (a) The Dorilton, 171 West 71st Street, New York City.

Reproduced with permission from the New York Public Library digital collections. (b) 127–131 W. 89 Street, New York City. Reproduced with permission from Avery Classics, Avery Architectural & Fine Arts Library, Columbia University.

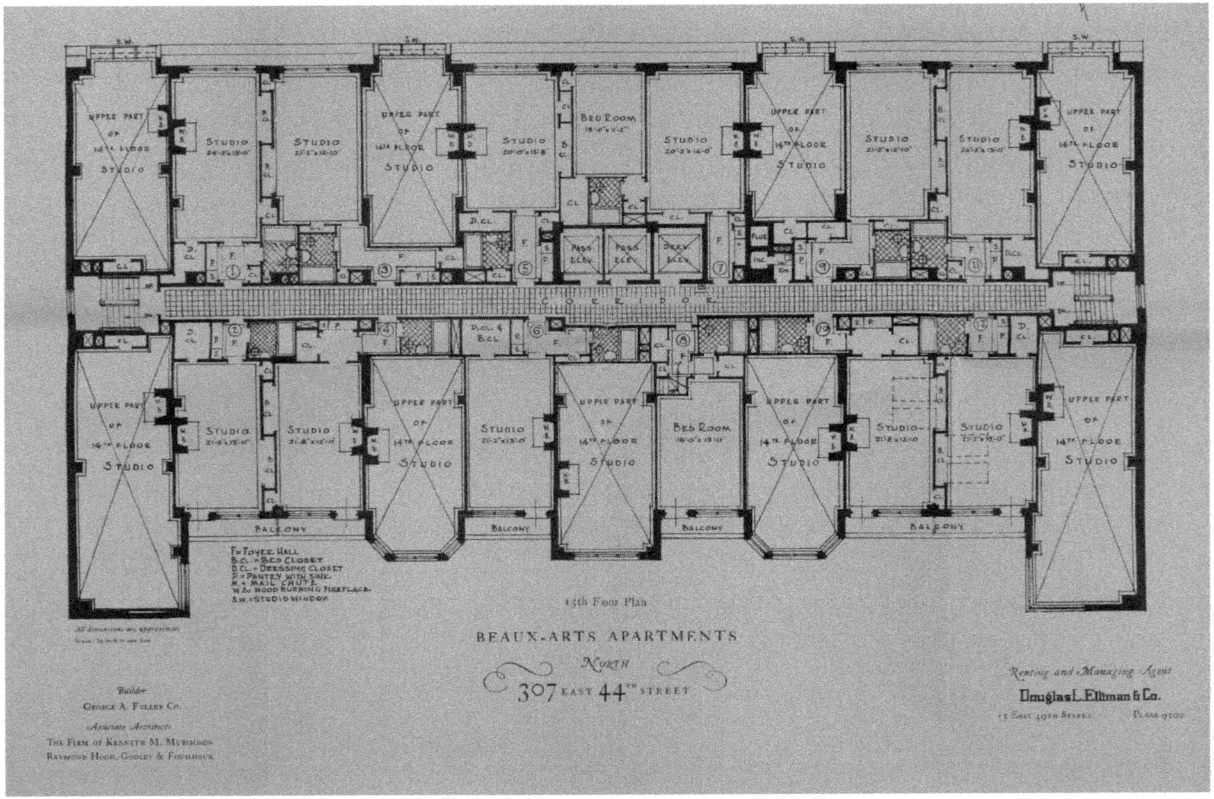

Figure 6.4 Double corridor apartment building, 307 E. 44 Street. Beaux Arts Apartments.

Reproduced with permission from Avery Classics, Avery Architectural & Fine Arts Library, Columbia University doi:10.7916/D8X650ZX

plans, significantly enhancing decision-making processes. It allows for forecasts regarding labour productivity [52], cost overruns [53], schedule delays, and building energy consumption, providing a strategic edge in managing complex projects.

BIM is a crucial platform for applying data science in construction, offering a comprehensive framework to capture, store, and manage extensive project data [54]. When combined with data science, BIM enhances both the design and operational phases of buildings by analysing large datasets generated throughout the project lifecycle. Techniques such as data mining are integral to BIM systems, further optimizing construction and operational efficiencies.

Despite its numerous advantages, the data-driven approach is not without its biases. Selection bias occurs when data used for analysis do not represent the entire population or project scope, potentially leading to skewed decisions [55]. For example, decisions based solely on data from urban projects can misguide rural construction planning. Sample adequacy bias arises from incomplete or insufficient datasets, yielding misleading insights. Out-of-group homogeneity bias assumes all projects or sites are similar, overlooking significant variations across locations, project types, or other factors [56]. For example, treating wood construction as a one-size-fits-all solution to reduce embodied carbon is a perfect example of out-of-group homogeneity bias derived from the Global North knowledge bias.

Methodological biases, including overgeneralization, correlation fallacy, confirmation bias, and automation bias, also play a role. Overgeneralization involves drawing broad conclusions from limited data without considering specific project contexts. The correlation fallacy mistakes correlation for causation, leading to erroneous assumptions. Confirmation bias interprets data in ways that reinforce pre-existing beliefs, ignoring contrary evidence. Automation bias results from over-reliance on automated systems without critically assessing their outputs.

One example of a correlation fallacy is the belief that merely installing solar panels on buildings will significantly reduce carbon emissions, without considering other essential factors like building orientation, energy efficiency measures, and overall energy consumption patterns. While there typically exists a correlation between the installation of solar panels and reduced electricity usage, this does not mean that solar panels directly cause reductions in energy usage. In some extreme cases, energy consumption may increase due to the rebound effect, where users consume more energy because it is perceived as 'free' once provided by solar panels.

Confirmation bias can also deepen environmental impacts. For instance, recent studies have championed timber for its low embodied carbon, often ignoring its higher costs and maintenance demands in regions where timber is scarce, like desert areas in the Global South. This bias can lead to decisions that overlook local material accessibility or long-term sustainability, potentially causing adverse environmental effects.

Societal and human biases further influence data-driven decisions. Historical biases reflect past discriminatory practices in the construction industry, such as gender or racial biases in hiring, which can be perpetuated through data-driven decisions. Cognitive biases like anchoring, framing effects, or status quo bias also affect decision-makers, influencing their interpretation and application of data-driven insights [57].

Finally, the reliance on data fosters a bias towards high-tech solutions that depend on continuous data monitoring and advanced technological infrastructure. This trend can marginalize traditional, low-tech methods that are often more sustainable and contextually appropriate. For example, high-tech solutions may optimize building energy use by adjusting mechanical systems, but they often neglect simpler methods like passive solar design, natural ventilation, or use of locally sourced materials that require no ongoing energy inputs or complex management systems.

6.2.3 Smart technologies and their embodied impact

Smart technologies, ranging from automated heating, ventilation, and air conditioning (HVAC) systems to intelligent lighting controls, are increasingly embedded within the built environment, offering enhanced comfort, greater energy efficiency, and lower operational costs. While these systems promise various benefits, their environmental impact—from production to disposal—might negate these advantages due to overlooked embodied impact. Haseeb et al. has revealed that the embodied carbon impact of smart sensors and other informational technologies—which includes their production, operation, and disposal—is substantial [58]. For instance, the electricity consumption from smart devices and related electronics accounted for 4% of the global total in 2007, increasing to 4.7% by 2012 [59]. Alarmingly, projections suggest that by 2030, electricity usage by smart technologies could be responsible for up to 23% of global carbon dioxide emissions [60]. This raises a critical question: with the increased adoption of smart technologies accompanied by increased energy and resource demand, how much environmental benefit can these technologies truly bring?

Smart technologies, despite their numerous benefits in promoting sustainability, can also have negative environmental impacts if not implemented or managed properly. One prominent issue is the proliferation of electronic waste (e-waste) resulting from the rapid turnover of smart devices and gadgets. The constant upgrading of technology and the short lifespan of many smart devices contribute to the generation of e-waste, posing significant challenges for proper disposal and recycling, which is often shipped to Global South developing countries, such as African countries, China, and India, as shown in Figure 6.5 [61]. According to the United Nations Global E-waste Monitor 2024 report, electronic waste (e.g. smartphones) has risen five times faster than documented e-waste recycling, a record of 62 million tonnes of e-waste was produced in 2022, up 82% from 2010, and the project is 82 million tonnes in 2030 [62]. In 2023, the leading countries exporting e-waste were the Netherlands, followed by France and Italy [63]. This e-waste is 'thrown' away from Global North developed countries to Global South. In a new report from CBS News, a horrific picture of mountains of garbage across acres of land, 'with tens of thousands of people sifting through it in places like the African national of Ghana' [64]. Disposal of e-waste like smartphones creates environmental and health dangers as many cell phone batteries contain arsenic, a highly poisonous substance, and other toxic materials like lead and mercury

Figure 6.5 (a) E-waste sorting in Ghana. Adapted from Flickr under a Creative Commons Attribution-ShareAlike https://www.flickr.com/photos/brsmeas/38272281662 (CC BY 2.0). (b) E-waste field in China. Adaption from Flickr https://www.flickr.com/photos/art_es_anna/415400769 (CC BY 2.0). (c) E-waste sorting in rural West Bengal, India. Adapted from Flickr https://www.flickr.com/photos/unevoc/50699047656 (CC BY 2.0).

[65]. Children as young as ten years old in Ghana sometimes live amid this toxic garbage and are desperate for a meagre payday [64].

Furthermore, the reliance on rare earth metals and minerals in the production of smart devices can have detrimental environmental consequences. One example given in Chapter 4 is bauxite mining. The extraction and processing of these materials, essential for manufacturing components of smart technologies, can result in habitat destruction, water pollution, and ecosystem degradation in mining areas, contributing to environmental harm [66]. And again, the production stage often happens in Global South countries that are far away from the end users, the consumers who live comfortably in Global North without having to bear those impacts.

Additionally, the energy consumption associated with smart technologies, particularly data centres and cloud computing infrastructure, can have a substantial carbon footprint. The continuous operation of data centres to support smart city initiatives, IoT devices, and digital services requires significant amounts of energy, often sourced from non-renewable fossil fuels, leading to increased greenhouse gas emissions and environmental degradation [67].

Besides environmental and energy concerns, the potential privacy and security risks associated with smart technologies, such as data breaches and cyberattacks, can have indirect negative environmental impacts. Incidents of data breaches not only compromise personal information but also lead to increased energy consumption for data recovery and security measures [68]. One notable example is the proposed Quayside project in Toronto, a partnership with Google's sister company Sidewalk Labs, which illustrated both the potential and pitfalls of such initiatives. The development plan aimed to create a high-tech neighbourhood, acting as a 'test bed' for new technologies [69]. Despite futuristic and sustainable claims, the project faced widespread criticism over data privacy concerns and a lack of transparency, highlighting the complex balance between innovation and public acceptance. Among the proposed smart technologies were 'smart mirrors' with built-in sensors and cameras to monitor health risks, which, despite their potential benefits for the elderly, raised significant privacy and ethical issues [70]. Ultimately, facing substantial criticism and financial viability challenges, by 2020, Sidewalk CEO announced that '[it had] become too difficult to make the project financially viable without sacrificing core parts of the plan' [71].

The Quayside project teaches us that while smart technologies present significant opportunities for enhancing sustainability in the built environment, they can also have negative environmental impacts if not carefully managed. Addressing issues like e-waste generation, energy consumption, resource extraction, and cybersecurity risks is essential to mitigate the adverse effects of smart technologies on the environment and promote their responsible and sustainable deployment.

6.3 Embracing low-tech strategies for sustainability

6.3.1 Is a high-tech solution necessary for reducing embodied carbon?

High-tech solutions such as digital twins, AI, and automation are frequently touted for their potential to reduce embodied carbon in construction. Yet, their real-world effectiveness is still under scrutiny due to unknown risks and ongoing debates about their immediate impact on carbon reduction. A closer examination of recent studies reveals a strong preference among practitioners and researchers for low-tech strategies, which are effective and practical for reducing embodied carbon emissions in construction projects. The author has recently conducted a study that explored six mitigation strategies for curbing the rise in embodied carbon in residential buildings: using space efficiently, extending building lifespans, reducing reliance on heavily processed materials, increasing use of wood products, enhancing recycling and reuse rates, and boosting use of prefabricated components [72]. The study found that using lighter-processed building materials, increasing wood usage, and reducing building size were the most effective for significant carbon reductions [72]. Notably, these solutions do not require advanced technologies.

A great example of a low-tech solution is material reuse, particularly celebrated within the circular economy for its significant role in reducing embodied carbon. Jayasinghe and Waldmann have demonstrated that reusing load-bearing elements in construction projects can significantly lower embodied energy and greenhouse gas emissions, contributing to a more sustainable and resource-efficient building sector [73]. Elements such as concrete or steel beams, partition walls, furniture, and carpets can all be reused and recycled to some extent. This approach does not require advanced technologies but rather effective regulations and a lifestyle shift.

Another low-tech solution is a structurally efficient design. Strategies such as reducing the volume of materials like asphalt and concrete and optimizing structural framing size show that sustainable practices do not always require cutting-edge technology. By optimizing building layouts, unnecessary transfer beams can be avoided, reducing material waste, carbon emissions, and costs. Additionally, designers can specify low-carbon materials and bio-based materials, significantly lowering a building's carbon footprint. These practices should be considered fundamental and essential rather than high-tech.

Building less and extending the service life of buildings are also effective strategies. Minimizing new construction and maximizing the use of current structures can significantly reduce the production of new materials and construction-related emissions [74]. Additionally, prolonging the use of existing structures minimizes waste generated from demolition and disposal of building materials, reducing environmental impact [75].

Overall, these strategies collectively illustrate that low-tech approaches are often more sustainable and contextually appropriate than high-tech counterparts. They highlight the importance of low-tech solutions that leverage traditional methods and natural materials, building less and extending the life of structures, proving that simpler, more traditional methods can provide substantial environmental benefits.

6.3.2 Why recycling and reuse are important?

Annually, humanity extracts roughly 100 billion tonnes of raw materials from the earth, equivalent to nearly two-thirds the mass of Mount Everest, as illustrated in Figure 6.6 [76]. Notably, about half of these raw materials are used in the construction sector, which contributes to a significant portion of global waste and at around 40% of the world's carbon dioxide emissions [76]. This environmental impact is substantial compared to the 2–3% of emissions attributed to the aviation industry, which often receives more public attention [76].

In construction and demolition (C&D), vast amounts of waste are generated through the new construction and the renovation or demolition of existing ones, including civil engineering deconstruction activities like roads, bridges, and dams. In 2018 alone, the U.S. generated 600 million tons of C&D debris—over twice the amount of municipal solid waste produced [77]. Interestingly, more than 90% of this debris comes from demolition activities, with less than 10% from new constructions, indicating a significant opportunity for recycling and reusing building materials and components [77].

Despite the belief that advanced technologies are essential for addressing issues in the built environment, recycling and reuse do not always require

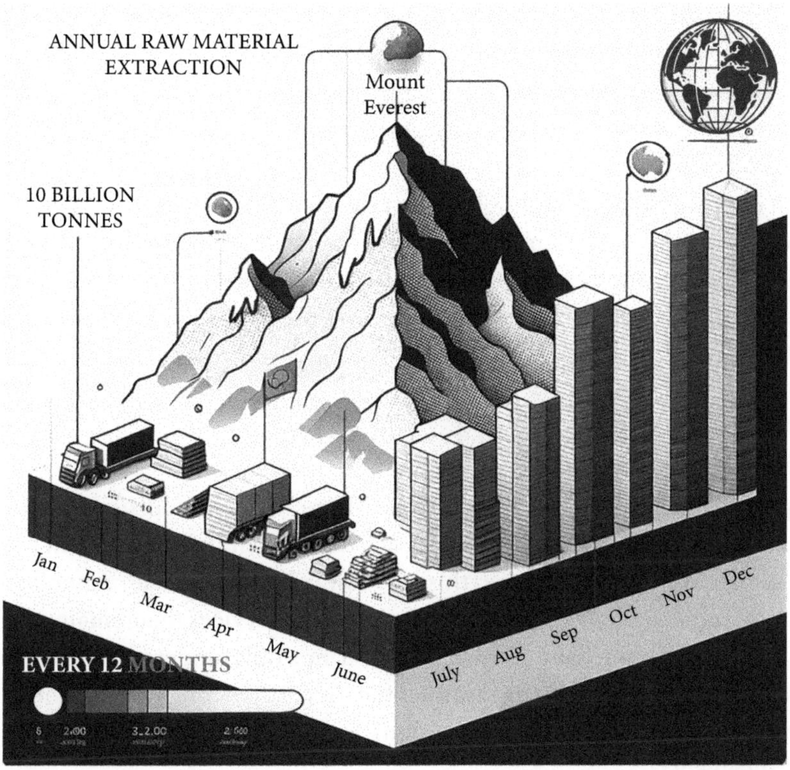

Figure 6.6 Global raw material extraction.

high-tech solutions. Experts point out that preparing existing materials for reuse is labour-intensive but simple, involving tasks like cleaning marble slab edges, removing mortar from tiles, and rewiring old electrical appliances. This process preserves jobs and creates new employment opportunities, challenging the notion that only machine automation and mass production represent advancement and development. In reality, manual labour and slower processes often signify a deeper level of craftsmanship and sustainability.

Moreover, the perception that recycling building materials is more costly than using new ones is often misguided. Although labour- and time-intensive, recycling can be more economical than hauling materials to landfills. Materials can be sorted on-site as they are removed or collected and transported to facilities for separation. Contractors can realize significant savings on landfill fees and even receive tax credits from government initiatives. For example, in Massachusetts, disposing of concrete blocks in a landfill costs about $140 per ton, whereas recycling the same material costs just $21 per ton [78]. Considering additional savings from recycling drywall, metal, and glass, the financial benefits can be substantial depending on the project's size. The U.S Environmental Protection Agency (EPA) has even developed a free Excel model to help calculate the feasibility and profitability of deconstructing buildings and recycling materials [79].

In conclusion, while technology is crucial in modern construction, it is not always the solution for environmental challenges. Sometimes, stepping away from a tech-centric approach and focusing on simpler, time-tested methods of recycling and reuse can offer more sustainable and cost-effective solutions.

6.4 Conclusion

A prevailing implicit bias towards the allure of novel tools can often lead decision-makers to embrace unnecessary risks, sidelining the prudent counsel of human expertise. Moreover, this bias towards cutting-edge technology can prompt substantial investments in products and services that remain untested or potentially hazardous. This inclination not only undermines established knowledge but also escalates the potential for costly errors in judgment, affecting the efficacy and safety of decisions in organizational and technological advancements.

The discussions within this chapter underscore the importance of integrating low-tech strategies alongside high-tech solutions to create a balanced approach to reducing embodied carbon and enhancing sustainability. Strategies such as material reuse, optimizing design and material quantities, and the use of lower carbon materials demonstrate that simplicity often aligns more closely with sustainability than complexity. These approaches not only preserve environmental integrity but also foster resilience and inclusivity in urban development.

For students and professionals in the built environment, this chapter serves as a call to evaluate the high-tech they employ critically. It encourages the adoption of a holistic perspective that values both innovation and tradition. As the field progresses, embracing a hybrid approach that combines the best of high- and low-tech solutions will be crucial in building sustainable futures that are not only technologically advanced but also deeply attuned to the ecological and social fabric of our communities.

Chapter 6 concludes Part II, which examines the concealed biases surrounding embodied carbon, unveiling their environmental justice implications and the influence of high-tech solutions. As we move forward, let us challenge the prevailing biases toward high-tech solutions by advocating for policies and practices that recognize the value of low-tech approaches. This balance is essential for truly sustainable development, ensuring that our pursuit of the new does not lead us to neglect the wisdom of the old. The path to sustainability is complex and requires a diverse toolkit—both old and new—to navigate effectively. In Part III, the author will explore the current global efforts to regulate embodied carbon, outline existing gaps, and propose a low-tech pathway to reduce embodied carbon by integrating knowledge from the Global South.

References

1. Elsbach KD, Stigliani I. New information technology and implicit bias. AMP. 2019 May; 33(2): 185–206.
2. Elsbach KD, Stigliani I. Evaluating new technology? You're more biased than you may realize. MIT Sloan Management Review [Internet]. 2020 Sep 23 [Accessed 16 May 2024]. Available from https://sloanreview.mit.edu/article/evaluating-new-technology-youre-more-biased-than-you-may-realize/
3. Bill Gates says AI will make it easier to combat climate change but must be 'used by people with good intent'. 2024 Jun 26 [Accessed 1 Jul 2024]. Available from https://news.sky.com/story/bill-gates-says-ai-will-make-it-easier-to-combat-climate-change-but-must-be-used-by-people-with-good-intent-13159556#:~:text=Bill%20Gates%20has%20said%20artificial,by%20people%20with%20good%20intent%22.
4. The White House. Fact sheet: the President's budget advances President Biden's unity agenda [Internet]. 2024 [Accessed 2024 May 9]. Available from https://www.whitehouse.gov/briefing-room/tatements-releases/2024/03/11/fact-sheet-the-presidents-budget-advances-president-bidens-unity-agenda/#:~:text=The%20Budget%20provides%20over%20%243,of%20the%20Administration's%20Executive%20Order.
5. Stam E, Wennberg K. The roles of R&D in new firm growth. Small Bus Econ. 2009 Jun; 33(1): 77–89.
6. Booltink LW, Saka-Helmhout A. The effects of R&D intensity and internationalization on the performance of non-high-tech SMEs. International Small Business Journal. 2018 Feb; 36(1): 81–103.
7. Li C, Li M, Zhang L, Li T, Ouyang H, Na S. Has the high-tech industry along the belt and road in China achieved green growth with technological innovation efficiency and environmental sustainability? IJERPH. 2019 Aug 27; 16(17): 3117.
8. Feng C, Hu Q, Kang Y, Li C, Zhang Y. The effects of Tech-Fin on corporate innovation: evidence from China. Accounting & Finance. 2023 Dec; 63(4): 3739–62.
9. Ulgiati S. A comprehensive energy and economic assessment of biofuels: when 'green' is not enough. Critical Reviews in Plant Sciences. 2001 Jan; 20(1): 71–106.
10. Beste A. Greenwashing & high tech! Faking it: (un-)sustainable solutions for agricultur [Internet]. 2021 [Accessed 24 Jun 2024]. (The Greens/EFA in the

European Parliament). Available from https://www.martin-haeusling.eu/images/GREENWASHING_AND_HIGH_TECH_-_Faking_it_un-sustainable_solutions_for_agriculture_WEB.pdf#page=2.18
11. Gates B. We need energy miracles [Internet]. 2014 [Accessed 9 May 2024]. Available from https://www.gatesnotes.com/energy-miracles
12. Varanasi A. You asked: does carbon capture technology actually work? [Internet]. 2019 [Accessed 9 May 2024]. Available from https://news.climate.columbia.edu/2019/09/27/carbon-capture-technology/
13. Klein J. As carbon capture heats up, old questions reignite [Internet]. 2023 [Accessed 9 May 2024]. Available from https://www.greenbiz.com/article/carbon-capture-heats-old-questions-reignite
14. Boot-Handford ME, Abanades JC, Anthony EJ, Blunt MJ, Brandani S, Mac Dowell N, et al. Carbon capture and storage update. Energy & Environmental Science. 2014; 7(1): 130–89.
15. Amelse JA, Behrens PK. Sequestering biomass for natural, efficient, and low-cost direct air capture of carbon dioxide (version 4) [Internet]. 2021 [Accessed 25 Jun 2024]. Available from https://www.preprints.org/manuscript/202106.0212/v4
16. Hamed AM, Tuan Kamaruddin TNA, Ramli N, Abdul Wahab MF. A review on blue and green hydrogen production process and their life cycle assessments. IOP Conf Ser: Earth Environ Sci. 2023 Dec 1; 1281(1): 012034.
17. Jacobson MZ. The health and climate impacts of carbon capture and direct air capture. Energy Environ Sci. 2019; 12(12): 3567–74.
18. IPCC. IPCC (Intergovernmental panel on climate change), special report: global warming of 1.51. [Internet]. 2018 [Accessed 25 Jun 2024]. Available from https://www.ipcc.ch/sr15/
19. Jacobson M. We don't need 'miracle' green technologies to save the planet [Internet]. 2023 [Accessed 9 May 2024]. Available from https://web.stanford.edu/group/efmh/jacobson/WWSNoMN/Oped-NewScientist.pdf
20. Haines A, McMichael AJ, Smith KR, Roberts I, Woodcock J, Markandya A, et al. Public health benefits of strategies to reduce greenhouse-gas emissions: overview and implications for policy makers. The Lancet. 2009 Dec; 374(9707): 2104–14.
21. Hu M, Zhang K, Nguyen Q, Tasdizen T. The effects of passive design on indoor thermal comfort and energy savings for residential buildings in hot climates: a systematic review. Urban Climate. 2023 May; 49: 101466.
22. West JJ, Smith SJ, Silva RA, Naik V, Zhang Y, Adelman Z, et al. Co-benefits of mitigating global greenhouse gas emissions for future air quality and human health. Nature Clim Change. 2013 Oct; 3(10): 885–9.
23. Remais JV, Hess JJ, Ebi KL, Markandya A, Balbus JM, Wilkinson P, et al. Estimating the health effects of greenhouse gas mitigation strategies: addressing parametric, model, and valuation challenges. Environmental Health Perspectives. 2014 May; 122(5): 447–55.
24. NUSBO. The Massachusetts affordable housing building challenge: passive housing new construction and retrofits at scale [Internet]. Northeastern University Sustainable Building Organization; 2022 [accessed 22 Apr 2022]. Available from https://web.northeastern.edu/nusbo/speakers_events/the-massachusetts-affordable-housing-building-challenge-passive-house-new-construction-and-retrofits-at-scale/

25. Northeast Sustainable Energy Association. Transforming affordable housing: mid-rise passive house retrofit with residents in place [Internet]. 2022 [Accessed 22 Apr 2022]. Available from https://nesea.org/session/transforming-affordable-housing-mid-rise-passive-house-retrofits-residents-place
26. ArchDaily. What is a traditional windcatcher? [Internet]. 2022 [Accessed 25 Jan 2022]. Available from https://www.archdaily.com/971216/what-is-a-traditional-
27. International Energy Agency. Buildings [Internet]. [Accessed 16 May 2024]. Available from https://www.iea.org/energy-system/buildings
28. U.S. Energy Information Administration. Use of energy explained [Internet]. [Accessed 16 My 2024]. Available from https://www.eia.gov/energyexplained/use-of-energy/electricity-use-in-homes.php#:~:text=Air%20conditioning%20use%20is%20now,1980%20to%2067%25%20in%202020.&text=The%20share%20of%20U.S.%20homes%20with%20central%20air%2Dconditioning%20systems,1980%20to%2067%25%20in%202020
29. U.S Energy Information Administration. How much electricity is used for air conditioning in the United States? [Internet]. 2024 [Accessed 26 Jun 2024]. Available from https://www.eia.gov/tools/faqs/faq.php?id=1174&t=1
30. Ghorbany S, Hu M, Yao S, Wang C, Nguyen QC, Yue X, et al. Examining the role of passive design indicators in energy burden reduction: Insights from a machine learning and deep learning approach. Building and Environment. 2023; 111126.
31. Hernández D. Understanding 'energy insecurity' and why it matters to health. Social Science & Medicine. 2016 Oct; 167: 1–10.
32. Yu Y, Kittner N. Hot or cold temperature disproportionately impacts U.S. energy burdens. Environ Res Lett. 2024 Jan 1; 19(1): 014079.
33. Venjakob M, Wagner O, Schnurr B. Can a CO_2 tax be socially just? Analysis of the social distribution effects of the German CO_2 taxation. Energies. 2023 Aug 28; 16(17): 6232.
34. ASHRAE. Improve energy efficiency in your home: cooling systems [Internet]. [Accessed 10 May 2024]. Available from https://www.ashrae.org/technical-resources/free-resources/summer-consumer-tips
35. Wang Z, Arlt ML, Zanocco C, Majumdar A, Rajagopal R. DeepSolar++: Understanding residential solar adoption trajectories with computer vision and technology diffusion models. Joule. 2022 Nov; 6(11): 2611–25.
36. Gartman M, Unger R, Golden R, Tang J, Shea R. What a 20 million heat pump commitment means for the US [Internet]. [Accessed 9 May 2024]. Available from https://rmi.org/what-a-20-million-heat-pump-commitment-means-for-the-us/
37. Edwards M, Garibay-Rodriguez J, Jacob Shimkus E, Shayan M, Shen X, Qiu Y, et al. Assessing inequities in electrification via heat pumps across the U.S [Internet]. [Accessed 9 May 2024]. Available from https://arxiv.org/pdf/2305.16488#page=1.53
38. Colding J, Wallhagen M, Sörqvist P, Marcus L, Hillman K, Samuelsson K, et al. Applying a systems perspective on the notion of the smart city. Smart Cities. 2020 May 20; 3(2): 420–9.
39. Dabaieh M, Maguid D, El-Mahdy D. Circularity in the new gravity—re-thinking vernacular architecture and circularity. Sustainability. 2021 Dec 29; 14(1): 328.
40. Fernandes JEP, Mateus R, Bragança L. The potential of vernacular materials to the sustainable building design. In: Vernacular Heritage and Earthen Architecture:

Contributions for Sustainable Development, edited by C Correia and R Correia. London: Taylor & Francis; 2014, pp. 623–9.

41. Hu M. Exploring low-carbon design and construction techniques: lessons from vernacular architecture. Climate. 2023; 11(8): 165.
42. Whyte W. The Englishness of English architecture: modernism and the making of a national international style, 1927–1957. Journal of British Studies. 2009; 48(2): 441–65.
43. Summary of the International Style. The Art Story [Internet]. [Accessed 12 May 2024]. Available from https://www.theartstory.org/movement/international-style/
44. Robin RT. Enclaves of America: the rhetoric of American political architecture abroad, 1900-1965. Vol. 145. Princeton, NJ: Princeton University Press; 2014.
45. Kornfield M. Two doors, few windows and 4,500 students: Architect quits over billionaire's mega dorm. The Washington Post [Internet]. 2021 Oct 30 [accessed 12 May 2024]. Available from https://www.washingtonpost.com/education/2021/10/30/ucsb-dorm-charlie-munger/
46. CBC Radio. Billionaire defends windowless dorm rooms for California students [Internet]. 2021 [Accessed 12 May 2024]. Available from https://www.cbc.ca/radio/asithappens/as-it-happens-the-tuesday-edition-1.6234150/billionaire-defends-windowless-dorm-rooms-for-california-students-1.6234462
47. Life inside a windowless hostel designed by billionaire Charlie Munger. Michigan News [Internet]. 2022 Apr 12 [Accessed 16 May 2024]. Available from https://darik.news/michigan/life-inside-a-windowless-hostel-designed-by-billionaire-charlie-munger/575960.html
48. Bethea C. Nightmare of the Windowless Dorm Room. 2021 Nov 13 [Accessed 2024 May 10]; Available from https://www.newyorker.com/magazine/2021/11/22/nightmare-of-the-windowless-dorm-room
49. Provost F, Fawcett T. Data science and its relationship to big data and data-driven decision making. Big Data. 2013 Mar; 1(1): 51–9.
50. Omole O, Ghosh S. Application of data science in the US commercial construction industry: current trends and future opportunities. International Journal of Construction Education and Research. 2024 Mar 23; 1–24.
51. K G. Big data in building information modeling research: survey and exploratory text mining. MOJCE [Internet]. 2017 Dec 19 [Accessed 14 May 2024]; 3(6). Available from https://medcraveonline.com/MOJCE/big-data-in-building-information-modeling-research-survey-and-exploratory-text-mining.html
52. Sadatnya A, Sadeghi N, Sabzekar S, Khanjani M, Tak AN, Taghaddos H. Machine learning for construction crew productivity prediction using daily work reports. Automation in Construction. 2023 Aug; 152: 104891.
53. Shoar S, Chileshe N, Edwards JD. Machine learning-aided engineering services' cost overruns prediction in high-rise residential building projects: Application of random forest regression. Journal of Building Engineering. 2022 Jun; 50: 104102.
54. Correa F. Is BIM big enough to take advantage of big data analytics? In Oulu, Finland; 2015 [Accessed 14 May 2024]. Available from http://www.iaarc.org/publications/2015_proceedings_of_the_32st_isarc_oulu_finland/is_bim_big_enough_to_take_advantage_of_big_data_analytics.html

55. Heckman J. Varieties of selection bias. The American Economic Review. 1990; 80(2): 313–18.
56. Akter S, McCarthy G, Sajib S, Michael K, Dwivedi YK, D'Ambra J, et al. Algorithmic bias in data-driven innovation in the age of AI. International Journal of Information Management. 2021; 60: 102387.
57. Daniel K. Thinking, fast and slow. London: Farrar, Straus and Giroux; 2011.
58. Haseeb A, Xia E, Saud S, Ahmad A, Khurshid H. Does information and communication technologies improve environmental quality in the era of globalization? An empirical analysis. Environ Sci Pollut Res. 2019 Mar; 26(9): 8594–608.
59. Galvin R. The ICT/electronics question: structural change and the rebound effect. Ecological Economics. 2015 Dec; 120: 23–31.
60. Andrae A, Edler T. On global electricity usage of communication technology: trends to 2030. Challenges. 2015 Apr 30; 6(1): 117–57.
61. Angelidou M, Psaltoglou A, Komninos N, Kakderi C, Tsarchopoulos P, Panori A. Enhancing sustainable urban development through smart city applications. JSTPM. 2018 Jul 12; 9(2): 146–69.
62. United National Institute for Training and Research. Global E-waste Monitor 2024: electronic waste rising five times faster than documented e-waste recycling [Internet]. [Accessed 1 Jul 2024]. Available from https://www.unitar.org/about/news-stories/press/global-e-waste-monitor-2024-electronic-waste-rising-five-times-faster-documented-e-waste-recycling
63. statista. Exports of electrical and electronic waste and scrap in selected countries worldwide in 2023 [Internet]. Available from https://www.statista.com/statistics/1389274/e-waste-exports-worldwide-by-country/
64. CBS NEWS. The tragic cost of e-waste and new efforts to recycle [Internet]. 2023 [Accessed 1 Jul 2024]. Available from https://www.cbsnews.com/news/the-tragic-cost-of-e-waste-and-new-efforts-to-recycle/
65. Marshall K. What are the dangers of disposing cell phones in landfills? [Internet]. 2014 [Accessed 1 Jul 2024]. Available from https://www.wcrecycler.com/blog/what-are-the-dangers-of-disposing-cell-phones-in-landfills
66. Olawumi TO, Chan DWM. Critical success factors for implementing building information modeling and sustainability practices in construction projects: a Delphi survey. Sustainable Development. 2019 Jul; 27(4): 587–602.
67. Attiya EE, Shebl MA, Magdy Darwish MTA. Integration of ecological and eco-technological design on architecture (hot, dry climatic environment). IOP Conf Ser: Earth Environ Sci. 2024 Jan 1; 1283(1): 012006.
68. Awuah KGB, Abdulai RT. Urban land and development management in a challenged developing world: an overview of new reflections. Land. 2022 Jan 14; 11(1): 129.
69. Leyland C. Surveillance capitalism': critic urges Toronto to abandon smart city project. The Guardian [Internet]. 2019 Jun 6 [Accessed 13 May 2024]. Available from https://www.theguardian.com/cities/2019/jun/06/toronto-smart-city-google-project-privacy-concerns
70. Urquhart M. Saint John-born writer delves into Google's failed attempt to build a smart city [Internet]. 2023 [Accessed 27 Jun 2024]. Available from https://www.cbc.ca/news/canada/new-brunswick/sidewalk-labs-google-failed-city-1.6933686

71. Jacobs K. Toronto wants to kill the smart city forever [Internet]. MIT Technology Review. 2022 [Accessed 27 Jun 2024]. Available from https://www.technologyreview.com/2022/06/29/1054005/toronto-kill-the-smart-city/
72. Hu M. Embodied carbon emissions of the residential building stock in the United States and the effectiveness of mitigation strategies. Climate. 2022 Sep 20; 10(10): 135.
73. Jayasinghe LB, Waldmann D. Development of a BIM-based web tool as a material and component bank for a sustainable construction industry. Sustainability. 2020 Feb 27; 12(5): 1766.
74. Hu M. A building life-cycle embodied performance index-the relationship between embodied energy, embodied carbon and environmental impact. Energies. 2020 Apr; 13(8).
75. Bertin I, Mesnil R, Jaeger JM, Feraille A, Le Roy R. A BIM-based framework and databank for reusing load-bearing structural elements. Sustainability. 2020 Apr 14; 12(8): 3147.
76. Miller N. The industry creating a third of the world's waste. 2021 Dec 15 [Accessed 2024 15j May]. Available from https://www.bbc.com/future/article/20211215-the-buildings-made-from-rubbish
77. U.S Environmental Protection Agency. Sustainable management of construction and demolition materials [Internet]. [Accessed 15 May 2024]. Available from https://www.epa.gov/smm/sustainable-management-construction-and-demolition-materials
78. Overstreet K. What does it cost to recycle building materials? Archdaily [Internet]. [Accessed 15 May 2024]. Available from https://www.archdaily.com/996805/what-does-it-cost-to-recycling-building-materials
79. U.S Environmental Protection Agency. Building material reuse and recycling estimating tool [Internet]. 2012. Available from https://www.epa.gov/land-revitalization/building-material-reuse-and-recycling-estimating-tool?utm_medium=website&utm_source=archdaily.com

PART III
ENVIRONMENTAL EQUALITY FOR THE FUTURE

Chapter 7
Embodied carbon policies and programmes

7.1 Three common carbon pricing mechanisms

Embodied carbon tax, carbon tariff, and carbon credits are all mechanisms designed to reduce carbon emissions and promote environmental sustainability. Each plays a unique role in addressing different aspects of the carbon emission problem [1]. An embodied carbon tax is a tax on the carbon emissions of products or services from their creation to disposal [1]. On the other hand, a carbon tariff is a border tax imposed on imported goods with high carbon emissions compared to domestically produced goods to prevent carbon leakage [2]. Carbon tariffs can be designed to tax carbon emissions directly and indirectly embedded in imports while providing rebates for exports from regions implementing carbon regulations [1]. While carbon tariffs and carbon taxes are policy-oriented measures, carbon credits operate within a market-oriented framework [3]. A carbon credit can impact trade structures, leading to changes in export and import patterns and affecting different industries in various ways [4]. In sections 7.1.1–7.1.4, we will first look into the three different policy strategies and frameworks; then we will have an overview of the existing policy landscape.

7.1.1 Embodied carbon tariff

Imagine you're running a race, but every time you get close to winning, some competitors take a shortcut and end up ahead of you. This is akin to what happens with carbon emissions in global trade. When some countries enforce strict climate rules, companies often move their operations to places with laxer regulations, where they can pollute more freely. This shift, known as 'carbon leakage', means that while one country becomes greener, another becomes dirtier, merely shifting pollution around. Those 'greener' countries and companies are essentially taking shortcuts by transferring their environmental burden to those with fewer resources to become greener, resulting in little overall benefit for the planet.

Embodied Carbon. Ming Hu, Oxford University Press. © Ming Hu (2025). DOI: 10.1093/9780198944485.003.0007

In the early 2000s, European countries began to worry about this problem. They were implementing ambitious climate policies and emissions trading schemes but feared their industries would suffer. The solution they devised was carbon tariffs—taxes on goods imported from countries with looser environmental rules, ensuring that the true cost of carbon emissions is reflected in their prices. The European Union (EU) took a leading role with its Emissions Trading System (ETS), launched in 2005 [5]. This cap-and-trade system limits overall emissions and allows companies to buy and sell allowances. As the EU tightened its climate targets, carbon tariffs emerged as a way to prevent companies from relocating to avoid stricter rules, thereby ensuring a level playing field.

However, this approach has not been universally welcomed. The unilateral implementation of carbon trading by the EU did not persuade other countries, particularly those in the Global South. Imagine a global summit where leaders discuss climate action. Carbon tariffs cause quite a stir. Leaders from developed countries in the Global North argue that these tariffs are necessary to meet global climate goals and prevent carbon leakage. In contrast, leaders from developing countries in the Global South point out that these tariffs could harm their economies and widen the gap between rich and poor nations. They argue that while they are committed to fighting climate change, they need support to do so without stifling economic growth or exacerbating poverty.

These concerns became more pronounced when, in October 2023, the EU approved the European Union's Carbon Border Adjustment Mechanism (CBAM), set to levy high-carbon goods imports, including steel and cement, starting in 2026. CBAM has already raised tensions with key trading partners, especially from the Global South. Countries like Brazil, South Africa, and India have accused the EU of discriminatory measures.

As an Indian government official put it, 'in the name of environment protection, EU is introducing a trade barrier that would hit not only Indian exports but also of many other developing countries' [6]. Consequently, India plans to file a complaint to the World Trade Organization over the EU's proposal to impose 20 to 35% tariff [6]. China has also strongly opposed this plan: in 2023, China called for a dedicated multilateral discussion at the World Trade Organization to enhance the understanding of controversial trade-related unilateral environmental measures such as CBAM.

Why are the Global South countries against carbon tariffs? They argue that these policies are inherently inconsistent and contradictory to multilateral climate and trade agreements, as well as with the EU's free trade agreement [7]. Although CBAM was framed as a global decarbonization policy, Global South countries believe it deviates from the burden-sharing principle under the Paris Agreement. CBAM makes no distinctions between or concessions for developing and least developed countries. Should EU countries pay back the profit gained by importing goods and materials from the Global South during the colonization period?

The Paris Agreement acknowledges that countries have common but differentiated responsibilities and capabilities, reflecting different national circumstances. Article 4, item 3, states,

> Each Party's successive nationally determined contribution will represent a progression beyond the Party's then current nationally determined contribution and reflect its highest possible ambition, reflecting its common but differentiated responsibilities and respective capabilities, in the light of different national circumstances. [8]

Furthermore, Article 4, item 4, emphasizes that

> Developed country Parties should continue taking the lead by undertaking economy-wide absolute emission reduction targets. Developing country Parties should continue enhancing their mitigation efforts, and are encouraged to move over time towards economy-wide emission reduction or limitation targets in the light of different national circumstances. [8]

Another contention is that revenue collected from CBAM will go to the EU budget instead of supporting capacity-building initiatives in developing countries, as required by the Paris Agreement. This deviation from the agreement's financial support obligations has led to strong pushback from the Global South. Article 9, item 1, of the Paris Agreement states,

> Developed country Parties shall provide financial resources to assist developing country Parties with respect to both mitigation and adaptation in continuation of their existing obligations under the Convention. [8]

Addressing the economic impact of these tariffs on developing countries is crucial. International trade often exacerbates inequality, highlighting the need for compensation mechanisms [9, 10]. By returning tariff revenues or providing financial support to affected countries, Global North nations can help mitigate the economic impact. The compensation hypothesis suggests that such measures can alleviate inequities and maintain public support for free trade.

Policies like emission taxes, refunding schemes, and entry-license taxes can also address these inequities [10].

Germany has taken a lead in implementing such policies. In 2022, it increased its international climate financing to a record €6.39 billion for climate protection and adaptation in developing countries. And Germany urges other Global North developed countries also 'pay their fair share of our joint promise to mobilize 100 billion US dollars per year to combat climate change in emerging and developing countries' [11]. As Federal Minister for Economic Affairs and Climate Action Dr. Robert Habeck pointed out, 'We can only curb global warming if all countries stick to their promises and steadily increase their climate protection efforts. We industrialised countries have committed ourselves to solidarity and support for developing and emerging countries. *Ultimately, it is about enabling the countries of the [G]lobal [S]outh to be able to meet their own climate protection obligation*' [11].

The success of carbon tariffs depends on balancing environmental goals with economic realities, ensuring climate action is both effective and fair. However, the effectiveness of such policies is debatable. A recent Asian Development Bank report states that CBAM is likely to reduce global carbon emissions by less than 0.2% while negatively impacting global exports and some EU manufacturers. If carbon tariffs are not the best solution, what other options do we have?

One alternative is implementing a carbon tax. Unlike carbon tariffs, a carbon tax imposes a direct fee on carbon emissions, encouraging businesses to reduce their carbon footprint without the complex international trade implications. In Section 7.1.2, we will explore how carbon taxes can provide a more straightforward and potentially more effective approach to mitigating climate change.

7.1.2 Embodied carbon tax

Carbon taxes are far from a radical concept; their origins lie in the early twentieth-century writings of economist Arthur Pigou [12]. Pigou proposed what is now known as a 'Pigouvian tax', a unitary tax imposed on polluters to make them bear the cost of the damage their emissions cause, which is often external to the polluter and invisible [13]. Although the actual damages from climate change are likely nonlinear, a carbon tax that scales linearly with carbon emissions effectively differentiates between various fossil fuels and activities based on their specific contributions to climate change.

Carbon tax and carbon tariff are distinct policy instruments aimed at addressing carbon emissions. A carbon tax is a fee imposed on companies based on the amount of carbon they emit, with the goal of reducing greenhouse gas emissions. On the other hand, a carbon tariff is a border tax imposed on imported goods with high carbon emissions compared to domestically produced goods. It aims to prevent carbon leakage and is often used as a political weapon. Unlike traditional carbon taxes that target direct emissions from energy consumption, an embodied carbon tax applies to the carbon footprint of products throughout their life cycle, from raw material extraction to manufacturing and transportation. This type of tax aims to incentivize businesses to reduce emissions across their supply chains by increasing the cost of carbon-intensive products. By internalizing the environmental cost of carbon emissions, the embodied carbon tax encourages more sustainable production practices and the use of low-carbon materials within the country and region's boundary.

There exist many different forms of carbon tax due to the variety of country and regional conditions. By 2020, thirty subnational, national, and supranational carbon tax initiatives were in place or scheduled for implementation [14]. Nordic countries are among the earliest to enact national carbon taxes. From 1990 to 1992, Norway, Sweden, Denmark, and Finland were the first to do so [13]. Even though these tax policies were progressive, the implementation of carbon taxes has been complicated by their membership in the EU, where rules prohibit pricing or taxing retail electricity in ways that discriminate against providers from other member states [15]. For example, Sweden has one of the world's highest carbon tax at around US$127 per ton of carbon dioxide [16, 17], but it exempts fossil fuels for electricity, and Finland has the same rule. This missed an opportunity to disincentivize the use of fossil fuels for generating electricity.

The second wave of carbon tax regulations began in 2008, with Switzerland and the Canadian province of British Columbia implementing carbon taxes in 2008, followed by Ireland (2010) [18], Iceland (2010) [19], Mexico (2012) [20], Japan (2012), France (2014) [21], and Chile (2014). Up to 2024, the carbon taxes have ranged from about US$0.61 (Indonesia) to US$167.17 (Uruguay) per ton of carbon emission [17]. Figure 7.1 shows the status of carbon tax initiatives on a global scale. What is worth noting is that a higher carbon tax does not directly translate into more carbon emission reduction or a more progressive policy. For example, Uruguay has the world's highest carbon tax, but this tax only covers about 5% of the country's total greenhouse gas emissions [22].

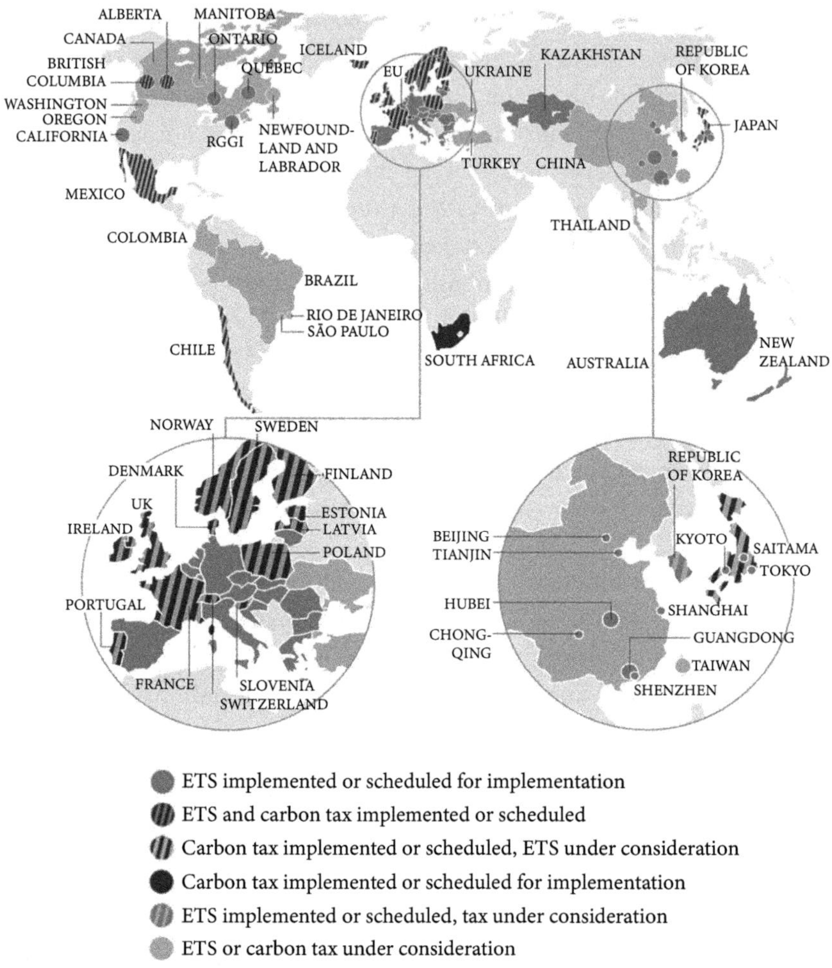

Figure 7.1 Global carbon tax status. Reproduced from Meike Naumann Visuelle Kommunikation; World Bank; Ecofys.

Vivid Economics via WikiMedia Commons (2016) under a Creative Commons 3.0 license https://commons.wikimedia.org/wiki/File:Carbon_pricing_initiatives_(2016).png (CC BY 3.0).

Different carbon tax policies vary in their tax rates, implementation methods, and impacts on income groups. Understanding these nuances is crucial for designing effective and equitable carbon emission strategies. Common types include fixed tax rate (FTR), progressive tax rate (PTR), revenue-neutral carbon tax (RNCT), indirect carbon tax (ICT), sector-specific carbon tax (SSCT), and hybrid carbon tax models (HCT) [13]. An FTR charges a fixed amount per ton of CO_2 emitted, providing price certainty and helping businesses plan their investments in emission-reduction technologies. PTR increases the tax rate over time or with emission levels, offering a stronger incentive for emitters to reduce their carbon footprint as emissions grow. RNCT offsets carbon tax revenues by reducing other taxes, such as income or corporate taxes, or by returning the revenue to the public through rebates or dividends, exemplified by British Columbia's carbon tax. This aims to maintain overall tax burdens while incentivizing carbon reduction. ICT targets goods and services contributing to carbon emissions rather than the emissions themselves, like energy taxes or taxes on carbon-intensive products. SSCT targets specific sectors, such as building or transport. For instance, South Korea's proposed carbon taxes for the building sector include direct taxes linked to property and energy taxes [23]. HCT combines elements of different taxes, such as gradually increasing the carbon tax rate while reducing other resource taxes and balancing economic impacts and environmental goals effectively.

A key question is how countries use their carbon tax revenue. Generally, there are three options: spending on environmental or climate change-related investments, returning money to taxpayers, or using it for general revenue purposes. Only five of the fifteen carbon-taxing countries allocate revenue to environmental or climate-related projects [17]. Japan and France spend 100% of their revenues on alternative energy and conservation measures [13]. Interestingly, early adopters like Nordic countries spend roughly half of their carbon tax revenue on general government use; in a way, those countries are actually putting the tax back into carbon generation. Understanding these different approaches helps in evaluating the effectiveness and equity of carbon tax policies.

7.1.3 Embodied carbon credits

The concept of carbon credits originated as part of global efforts to combat climate change, particularly through mechanisms established under the Kyoto Protocol [24]. Unlike carbon taxes and tariffs, carbon credits and carbon trading programmes are typically considered environmental policies rather than tax policies [25]. The definition of carbon credit is listed in Box 7.1.

Box 7.1 Definition of carbon credit

Carbon credits are permits or certificates that represent the right to emit one metric ton of carbon dioxide or an equivalent amount of other greenhouse gases. These credits are part of cap-and-trade systems, where a government or regulatory body sets a limit (cap) on the total amount of emissions allowed within a specific period. Companies that reduce their emissions below the cap can sell their excess allowances as carbon credits to other companies that need to offset their emissions.

Two main mechanisms under the Kyoto Protocol that facilitated the creation and trading of carbon credits are the Clean Development Mechanism (CDM) and Joint Implementation (JI) [24]. The CDM enables developed countries in the Global North to invest in emission-reduction projects in developing countries in the Global South, earning certified emission-reduction (CER) credits that can be used to meet their own targets. This promotes sustainable development and technology transfer to developing countries. Similarly, JI allows countries to earn emission-reduction units (ERUs) from projects in other developed countries, encouraging collaboration between nations with established emissions targets under the Kyoto Protocol.

The introduction of these mechanisms began in the EU, especially after the United States withdrew from the Kyoto Protocol. The EU then drafted and passed the Emission Trading System (ETS) after over a year of negotiations among the European Parliament, industry stakeholders, and politicians, leading to its adoption in 2003 and implementation starting in 2005 [26]. The ETS gained support from various industry lobby groups, financiers, and NGOs within EU, forming a powerful coalition. This was notable because many of these groups had opposed the original EU carbon tax proposal [27]. *Unlike a carbon tax, which would have involved significant financial transfers from industry to the government, the ETS avoided this by allocating emissions allowances to polluters for free* [3]. A key feature of the ETS was that it allowed national governments to make decisions on allowance allocations. This approach reduced concerns about granting new fiscal powers to the EU and aligned with the principle of 'subsidiarity', allowing Member States to manage allocations internally [3].

This market-based approach provides financial incentives for businesses to invest in emission-reduction technologies and practices. Additionally, carbon credits can be generated through projects that remove or reduce greenhouse gases from the atmosphere, such as reforestation, renewable energy installations, energy efficiency improvements, and carbon capture. It seems like a very reasonable system. However, carbon credits have faced several criticisms and unintended consequences, highlighting their complex role in the fight against climate change. One major concern is over-crediting, where projects receive more credits than the actual climate benefits they provide [28, 29]. Over-crediting can be directly caused by double counting, especially in voluntary markets with less stringent regulations. In these scenarios, the same emissions reductions can be counted by both the seller and the buyer of the credits, distorting the actual environmental benefits and misleading stakeholders about the effectiveness of carbon offsetting measures [30]. This practice undermines the integrity of carbon offset programmes and devalues legitimate climate change mitigation efforts [31].

Another significant issue is the quality and integrity of the credits themselves, leading to the issuance of invalid credits, often referred to as junk carbon credits [32]. For instance, Verra, one of the largest carbon credit certifiers, has been criticized for approving offsets that are considered 'worthless' by some experts. Investigations have revealed issues such as overstated benefits from rainforest protection projects and questionable accounting practices in initiatives like rice cultivation projects. These problems highlight the risk that some carbon credits may not genuinely represent real emission reductions, undermining the entire system's credibility [30]. Furthermore, some carbon credit projects' lack of stringent monitoring and verification processes has raised doubts about the actual emission reductions achieved [33].

Moreover, carbon credits have been criticized for allowing wealthy countries and corporations to continue polluting while purchasing relatively cheap offsets. This practice overlooks the social and economic impacts on local communities in countries where the pollution occurs. For example, the rapid implementation of REDD+ projects (Reducing Emissions from Deforestation and Forest Degradation) has sometimes led to conflicts with local communities, who may lose access to their land and resources without adequate compensation or involvement in decision-making processes [30]. This can reduce the incentive for these entities to make substantial reductions in their own carbon footprints. Critics argue that this system effectively lets polluters buy their way out of making necessary and more challenging changes to their operations and consumption patterns [30].

The plastic credit is one example. In 2023, the United Nations met in Nairobi to draft a global treaty on plastic pollution. The conglomerates responsible for much of the waste lobbyied for 'innovative' measures like the growing offsetting market to be part of the solution. Instead, it's fast becoming part of the problem, allowing polluters to buy their way out of trouble. As described by the science and policy director at Global Alliance for Incinerator Alternatives: 'The big challenge is we need to make less plastic, but the whole point of plastic credit is it gives plastic companies a "Get Out of Jail Free" card.' [34] What makes this situation worse is that such credits allow large corporations to continue greenwashing. For instance, Nestlé Philippines has spent $732,900 on credits, all for burning plastic and none for recycling, releasing an estimated 16,544 tonnes of carbon dioxide, and has been certified by PCX as 'plastic net zero'. Similarly, PepsiCo Philippines' credit purchases have all gone towards projects where plastic was burned [34].

Lastly, the complexity of carbon credit systems and the reliance on simulation models for credit generation also pose challenges. For instance, soil organic carbon simulation models are under pressure to provide a reliable foundation for carbon credit generation [35]. Additionally, accurate assessment to control deforestation and degradation rate errors is crucial for ensuring the effectiveness of carbon credit programmes [36].

7.1.4 Comparison and conclusion

As demonstrated in Figure 7.2, carbon credits operate on a global scale to limit maximum carbon emissions and regulate carbon trade. In contrast, carbon tariffs function only among international trading partners, giving them a more limited reach. Lastly, carbon taxes are domestic policies that function within a country's borders. Each mechanism offers unique advantages and poses distinct challenges, which must be carefully considered to develop effective and equitable climate policies.

Carbon taxes stand out for their simplicity and predictability. By imposing a clear cost on carbon emissions, they encourage businesses and consumers to adopt more sustainable practices. Additionally, the revenue generated from carbon taxes can be strategically reinvested in renewable energy projects or returned to the public, potentially offsetting the financial impact on households. However, there are downsides. Higher production costs can be passed on to consumers, increasing the cost of living. Moreover, without complementary measures like carbon tariffs, domestic industries might

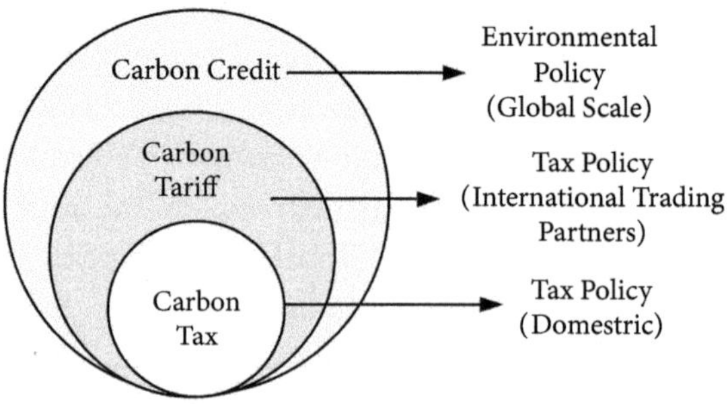

Figure 7.2 Carbon credit, carbon tariff, carbon tax.

face competitive disadvantages against businesses in countries with fewer environmental regulations.

Building on this, carbon tariffs provide a solution to the issue of carbon leakage, where companies relocate production to countries with laxer climate policies. By taxing imported goods based on their carbon footprint, carbon tariffs encourage global industries to adopt greener practices to remain competitive. This not only protects domestic industries but also promotes global climate action. However, the implementation of carbon tariffs can lead to trade disputes and place an economic strain on developing countries in the Global South, which have inherent disadvantages that result in less competitive products. Furthermore, calculating and enforcing these tariffs can be complex and resource-intensive, requiring an accurate assessment of the carbon content of imported goods.

Transitioning from policy to market-based solutions, carbon credits offer a flexible approach, allowing businesses to offset their emissions by purchasing credits from projects that reduce or sequester carbon elsewhere. This system incentivizes innovative emission-reduction technologies and fosters international cooperation, particularly through projects in the Global South developing countries.

It is essential to leverage each mechanism's strengths while addressing their drawbacks to develop a robust and effective embodied carbon policy. Synergistic policies include using carbon taxes to generate revenue for renewable energy projects, applying carbon tariffs to prevent carbon leakage and protect domestic industries, and ensuring equitable standards for fair carbon emission allowance allocation using carbon credit systems. By integrating these mechanisms into a cohesive strategy, policymakers and markets can create a more balanced approach that promotes sustainability, economic equity, and environmental protection. But what does the reality look like? Section 7.2 reveals the current landscape of embodied carbon-related programmes and policies.

7.2 Embodied carbon policies and programmes

7.2.1 Global landscape of embodied carbon policies

According to the Carbon Leadership Forum's Embodied Carbon Policy Toolkit, as of 2023 there are 186 embodied carbon-related policies across twenty countries. This count excludes carbon taxes and tariffs. These

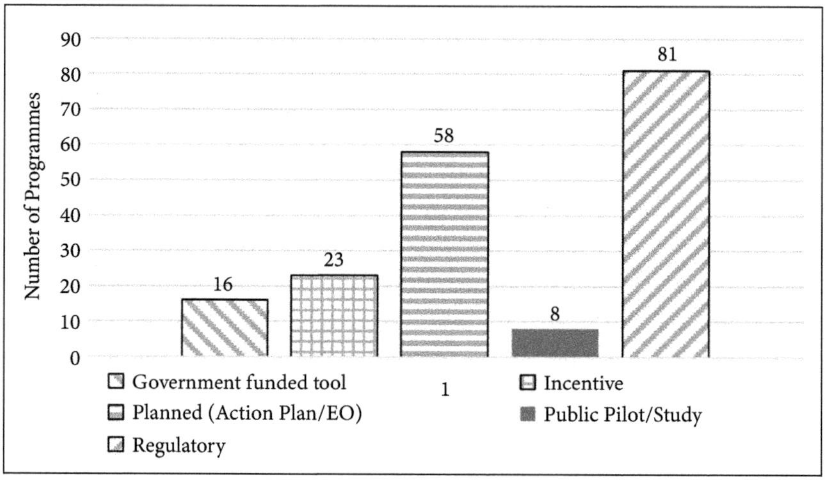

Figure 7.3 Types of programmes.

initiatives highlight a significant global intention to address carbon emissions embedded in goods and services throughout their life cycles beyond just direct emissions. As illustrated in Figure 7.3, those policies can be organized into five categories: regulatory, planned (action plan/Executive Order), incentive, government-funded tool, and public policy/study. Regulatory is the largest category, with 81 total. Regulation has a variety of formats; some prescribe how the new construction should be built, and others set the limit of the total embodied carbon allowed from a building from the life-cycle measurement. For example, on 1 January 2023, an addendum to the Danish Building Regulations came into force, stipulating that a life-cycle assessment must be conducted on all new buildings to calculate their environmental impact. A maximum global warming potential (GWP) threshold of 12 kg CO_2 equivalent per square metre per year must also be documented on buildings larger than 1,000 m^2 [37]. In France, starting in 2022, all new public buildings are mandated to be built from at least 50% timber or other natural materials. In addition to wood, the mandate can be fulfilled by bio-based materials made from matter derived from living organisms, such as hemp and straw [38].

The second largest category is planned (action plan/Executive Order), 59 programmes. For example, Greening Government Strategy: A Government of Canada Directive requires the Canadian government to disclose the amount of embodied carbon in the structural materials of major construction projects by 2022 based on material carbon intensity or a life-cycle analysis. The government must reduce the embodied carbon of the structural materials of major construction projects by 30%, starting in 2025, using recycled and lower-carbon materials, to comply with performance-based design standards. The government must conduct whole building (or asset) life-cycle assessments by 2025 at the latest for major buildings and infrastructure projects [39]. Another action plan example is the City of Helsinki's Roadmap for Circular and Sharing Economy [40]. This action plan aims to achieve a carbon-neutral circular economy by 2050. The roadmap focuses on sustainable construction, including preserving and repurposing existing buildings, integrating low-carbon construction practices from pre-construction to project completion, and exploring alternatives to concrete, such as low-emission concrete or substitutes, to mitigate carbon impact. The roadmap emphasizes the adoption of circular economy principles in new construction and renovation projects, with a focus on criteria like efficient use of materials, adaptability, modularity, use of recycled materials, and ease of repair and reusability, along with calculating the life-cycle costs and carbon footprint in construction projects [41].

The third category with the largest programmes is incentives. Those incentives exist in the U.S., Austria, Canada, Denmark, New Zealand, Switzerland, the U.K., and Sweden at the city, county, and national levels. As an incentive, the local jurisdiction often rewards the developer or builder for constructing low-embodied carbon buildings with more allowed buildable areas or reduced project costs. For example, the Quebec Wood Charter is a policy that encourages the use of wood for all publicly funded infrastructure projects [42]. As a result of this policy, the provincial building code was updated to increase the maximum height for wood structures from four stories to six stories and taller. Douro-Dummer's innovative Sustainable Development Program was launched in early 2020 as Canada's first city-wide embodied carbon reduction programme for building construction. It provides, at this time, a 40% permit fee rebate on all approved projects that meet the required greenhouse gas reduction targets or an 80% permit fee rebate combined with net-zero-ready construction [43].

For the fourth category, government-funded tools and databases are mostly at the national level. For example, Finland and the Netherlands have a single national methodology for the whole-life-cycle assessment of buildings. Other countries, such as Austria, Belgium, and France, have their own environmental production declaration (EPD) programmes and databases for construction materials and products.

When we look into the programme distribution by country, as shown in Figure 7.4 (left), the U.S. takes the significant lead, and it is worth noting that

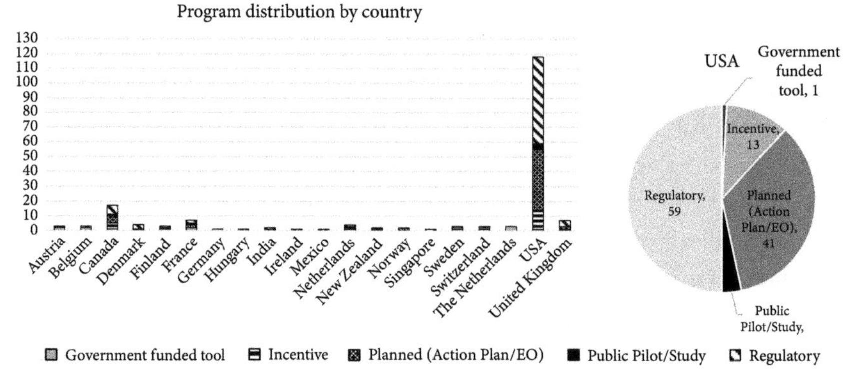

Figure 7.4 (left) Programme distribution by country; (right) programme distribution in the U.S. by percentage.

most of the programmes are regulatory (Figure 7.4 right). What is interesting is that even the U.S. seems to be taking the lead in regulating embodied carbon emissions; however, those policies and programmes are mainly produced in thirteen cities, which indicates a clearly uneven distribution of the policies, only covering less than 1% of jurisdiction in the U.S. We can also interpret it as the lack of national consensus for the requirements and policy. Moreover, the effectiveness of those policies and programmes is still being determined and undocumented.

When we compare the Global North and South, the disparity is very apparent. As shown in Figure 7.5, only three countries in the Global South have programmes: India, Mexico, and China. Instead of immediately blaming the Global South for its inaction, the real question we should ask is whether those Global North-led policies and approaches are suitable, fair, and practical for the Global South.

In Chapter 4, we explained why consumption-based carbon emissions are better measures than production-based emissions. Figure 7.6 illustrates the correlation between consumption-based CO_2 emissions per capita and GDP per capita. GDP per capita is often used to measure the living standard. The result is clear: regardless of the size of the country, wealthy Global North developed countries have higher per-person carbon emissions than those in poorer Global South developing countries. For example, Qatar, a small but very wealthy nation, has over twenty times more per capita carbon emissions than Nigeria. Similarly, China's neighbouring countries Japan and South Korea—both considered Global North developed nations—also exhibit much higher per capita carbon emissions. The policies, standards,

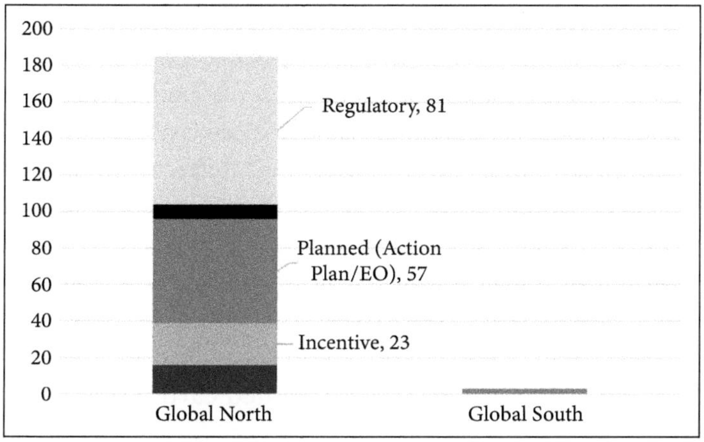

Figure 7.5 Global North and Global South embodied policy distribution.

and targets established by Global North countries are often not well received by Global South countries. For example, an overall emission cut measured by percentage might not be practical for Global South countries with low

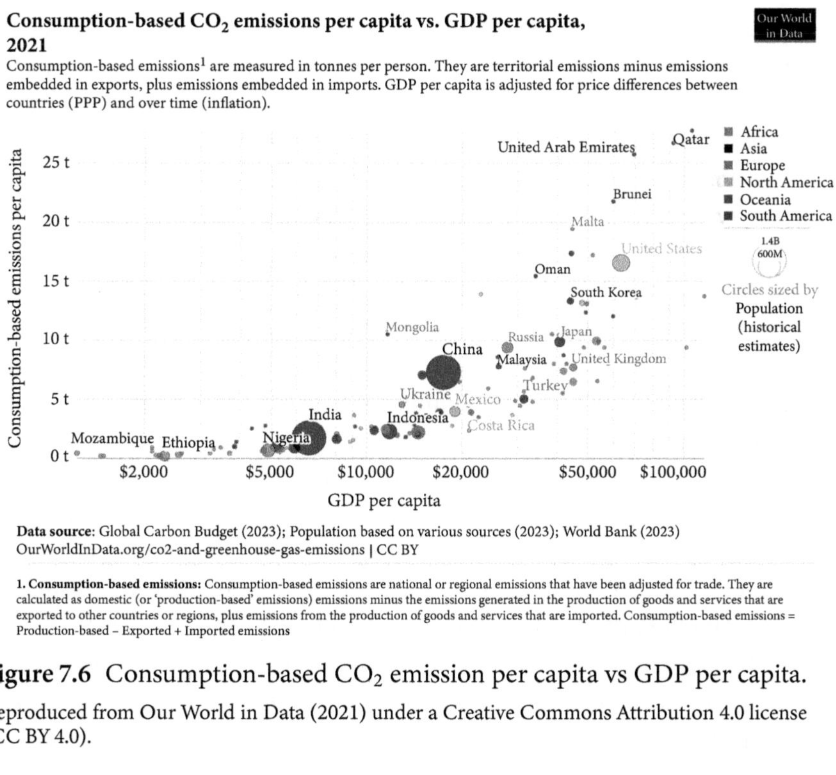

Figure 7.6 Consumption-based CO_2 emission per capita vs GDP per capita.

Reproduced from Our World in Data (2021) under a Creative Commons Attribution 4.0 license (CC BY 4.0).

GDP per capita because these countries need to mitigate carbon emissions while continuing to grow their economies to achieve equity in human development until the GDP per capita reaches a level similar to those of Global North. As their GDP per capita increases, it's likely that their CO_2 emissions per capita will also rise. The significant differences in economic status and carbon emissions between Global North and South underscore the need for more tailored and equitable approaches to global carbon reduction efforts.

7.2.2 Whole building embodied carbon limits

All policies and programmes aiming to mitigate embodied carbon are beneficial, but the most effective strategy involves setting strict limits on the maximum amount of embodied carbon allowed, similar to how operational carbon is regulated. Among the 185 policies reviewed, only 19 programmes established specific limits for either whole building embodied carbon requirements or specific building materials, such as concrete. The remaining programmes focus on providing high-level guidelines. These 19 programmes originate from five countries and show significant variations in their limits. As detailed in Table 7.1, the limits range from 250 to 970 kg CO_2e/m^2, with a median value of 500 kg CO_2e/m^2, for fifty years.

One interesting city-wide policy to look into is Zurich. Zurich committed to the vision of the 2000 Watt Society and has set a 2050 target of 8.5 kg CO_2e/m^2/year (425 kg CO_2e/m^2 for fifty years) for life-cycle embodied carbon in residential buildings. The 2000-Watt Society is a concept introduced in 1998 by the Swiss Federal Institute of Technology in Zurich (ETH Zurich) that aims to reduce the average primary energy use of Global North countries' citizens to no more than 2,000 watts (equivalent to 2 kW-hr per hour or 48 kW-hr per day) by 2050, without compromising their standard of living [44]. Is this goal achievable?

As of 2023, Switzerland's energy use per capita stands at 35.77 kilowatt-hr (kWh) per hour, reflecting a 15% increase since 1965, and still far from the target of 2 kWh per hour. In contrast, Bangladesh has the lowest energy consumption at 2.94 kWh per hour, while Qatar has the highest at 205.09 kWh per hour. All Global North countries use significantly more energy than those in the Global South. This raises a philosophical question: can the lifestyle supported by 35.77 kWh per hour possibly be maintained if energy use is reduced to 2 kWh per hour? The answers remain unknown.

Table 7.1 Whole building life cycle embodied carbon policy

Country	City	Requires wbLCA or embodied carbon calculation	Requires Type III EPD	Embodied carbon limit (building) ($kgCO_2e/m^2$)	Embodied carbon limit (concrete) ($kgCO_2e/m^2$)	Embodied carbon limit (steel) ($kgCO_2e/m^2$)
U.S.	Pittsburgh	Yes	No	—	—	—
	Austin	Yes	Yes	500	—	—
	Somerville	Yes	Unknown	350	—	—
	Cambridge	Yes	No	—	—	—
	Kirkland	Yes	No	Varies by project	—	—
	Lexington	Yes	No	—	—	—
	San Diego	No	No	—	—	—
	Seattle	No	Yes	—	—	—
	Berkeley					
	Marin County (CA)	No	Yes	—	Varies by volume and strength	—
	Denver	No	Yes	—	Varies by volume and strength	Varies by volume and strength
	New York City	No	Yes	—	—	—
	Portland	No	Yes	—	Varies by volume and strength	—
Switzerland	Zurich	Yes	No	425 (residential)	—	—
Denmark	Nation-wide	Yes	—	600	—	—

Continued

Table 7.1 *Continued*

Country	City	Requires wbLCA or embodied carbon calculation	Requires Type III EPD	Embodied carbon limit (building) ($kgCO_2e/m^2$)	Embodied carbon limit (concrete) ($kgCO_2e/m^2$)	Embodied carbon limit (steel) ($kgCO_2e/m^2$)
U.K.	London	Yes	No	970 (Office)	—	—
		Yes	No	800 (Residential)	—	—
		Yes	No	675 (School)	—	—
		Yes	No	690 (Retail)	—	—
Canada	Douro-Dummer, ON	Yes	No	Varies by project	—	—
	Vancouver	Yes	No	400	—	—
	Toronto	Yes	No	250–350 (mid-high rise & non-residential)	—	—
		Yes	No	<250 (low-rise residential)	—	—
		Yes	No	<350 (city-owned facility)	—	—
	Langford, BC	No	Yes	—	Varies by compressive strength	—

Note: Denmark's limit is 12 $kgCO_2e/m^2$. It is then calculated based on the fifty-year life span of the building to obtain 600 $kgCO_2e/m^2$.

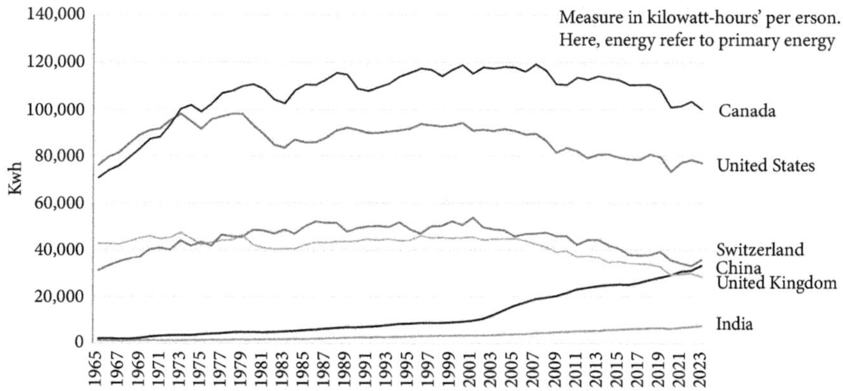

Figure 7.7 Energy use per person.

Source: data from U.S. Energy Information Administration (2023) Energy Institute—Statistical Review of World Energy (2024); population-based from various sources (2023).

Figure 7.7 compares energy use per capita across major Global North and South countries, with Switzerland as a reference point. The figure tells a compelling story of energy consumption over the decades, measured in kilowatt-hours per person. Canada and the U.S. stand out with their high energy consumption per capita, peaking in the early 2000s. Both countries have since embarked on a downward trend, with the U.S. starting its decline around 2005, followed closely by Canada. In contrast, Switzerland and the United Kingdom exhibit moderation and stability, maintaining relatively steady and much lower energy use per capita than the U.S. and Canada, although they have seen a gentle decline in recent years. The contrast between Canada and the U.S., the U.K., and Switzerland demonstrates that similar living standards can be met at low energy use intensity, and with increased efficiency, energy use can be lowered further.

On the other hand, China's trajectory shows a rapid increase in energy consumption starting in the late 1990s, reflecting its industrial rise and economic boom. Meanwhile, India's energy use per capita, though starting from the lowest point, has been steadily and significantly increasing, especially since the late 2000s. This dynamic picture highlights a broader theme: while Global North developed nations have stabilized their energy use intensity, there is still considerable progress needed to reduce consumption and emissions to meet their targets, particularly for countries like the U.S. and Canada. Even in Switzerland, current energy use intensity is nearly 18 times higher than desired.

7.2.3 Global South progress

As we've seen, most embodied carbon-related policies are concentrated in the Global North, while the Global South is just starting to recognize and address these issues since economic growth is still their priority. Take China, for example. While the central government issues broad guidelines, specific policies are crafted at the provincial and local levels. This decentralized approach, much like the system in the U.S., allows regions to tailor their policies to local needs, balancing economic growth with efforts to cut carbon emissions. But this also means navigating a complex web of regulations.

China made an early start with its GB/T 50378–2014 standard, which introduced operational carbon calculations per unit of building area and paved the way for carbon assessments in embodied carbon. This standard uses a scoring system based on seven performance categories, awarding buildings an overall rating between 0 and 3 stars, with at least forty points needed for a '1-star' rating. The updated GB/T 50378–2019 standard placed even more emphasis on carbon performance [45].

Despite these steps, the uptake has been less than ideal. In China, over 65% of survey respondents admit they don't measure carbon emissions during a project's operational period. Similarly, 66% don't track emissions from construction materials [45]. The main hurdle? A lack of harmonized standards and tools for carbon measurement is a problem cited by nearly half of the respondents. Clearly, there's a pressing need for better methodologies and instruments, a challenge that's also familiar in the United States.

Now, let's look at India, where the potential for carbon savings lies in the future. About 70% of the urban infrastructure India will need by 2030—everything from drinking water and sanitation to electricity, gas distribution, and transport—has yet to be built. This makes it crucial to minimize the carbon footprint of these upcoming projects [46]. In India, the built environment is responsible for both embodied and operational carbon emissions, accounting for 40 and 60%, respectively [46]. This high share is largely due to the extensive use of carbon-intensive materials like concrete and steel. Consequently, India is focusing on regulating the embodied carbon of these materials. Two significant initiatives stand out.

First, the Indian national standard introduced in 2023 (IS 18189) limits the use of Portland cement by advocating for low-carbon alternatives. For instance, LC3-50 cement blend, which consists of 30% calcined clay, 15% limestone, 5% gypsum, and 50% clinker, reduces carbon emissions by about 40% compared to traditional Portland cement concrete and can cut overall costs by up to 25% [47]. Second, India is establishing a national database of embodied carbon in building materials. This effort, spearheaded by the Center for Advanced Research in Building Science and Energy, is supported by the Ministry of New and Renewable Energy and the Gujarat Energy Development Agency. These measures are pivotal for India's sustainable development and in reducing the carbon footprint of its future infrastructure [48].

So, while the Global South faces unique challenges, the efforts in China and India highlight the critical steps and substantial potential for reducing embodied carbon in the built environment. By addressing these challenges with tailored and innovative solutions, these countries can pave the way for a greener, more sustainable future.

7.3 Conclusion

The chapter has explored three critical mechanisms for addressing carbon emissions: embodied carbon tax, carbon tariff, and carbon credits. Each mechanism offers distinct advantages and challenges in the global effort to reduce greenhouse gas emissions and promote environmental sustainability. The embodied carbon tax incentivizes businesses to reduce emissions across their supply chains by internalizing the environmental costs, encouraging the adoption of more sustainable practices. Carbon tariffs address the issue of carbon leakage by imposing border taxes on imported goods with high carbon emissions, thus promoting global adherence to greener practices. Meanwhile, carbon credits offer a flexible market-based approach, providing financial incentives for businesses to invest in emission-reduction technologies and offset projects. All three mechanisms have unintended consequences and loopholes that can exacerbate Global North and South disparity. Effective embodied carbon policy should integrate these tools to harness their benefits while mitigating their limitations.

Besides the three carbon pricing control mechanisms, the global landscape of embodied carbon policies reflects a significant effort to address carbon emissions, with 186 policies across twenty countries. However, there is a noticeable disparity between the Global North and South, highlighting the need for fair and practical policies that consider the unique circumstances of developing regions. Programmes that set specific limits on whole building embodied carbon, such as those in Zurich, provide valuable benchmarks and demonstrate the potential for rigorous, localized approaches to carbon reduction.

In conclusion, a balanced and integrated policy framework that incorporates carbon taxes, tariffs, and credits and creates embodied carbon limits at the whole building level is essential for advancing sustainable development and achieving meaningful progress in reducing carbon emissions. By considering the pros and cons of each mechanism and implementing them in a complementary fashion, policymakers can create a more resilient and equitable path toward a sustainable future.

References

1. Böhringer C, Bye B, Fæhn T, Rosendahl KE. Alternative designs for tariffs on embodied carbon: a global cost-effectiveness analysis. Energy Economics. 2012 Dec; 34: S143–53.
2. Shuai C, Guo Q, Ding L, Cheng X. Modelling carbon tariffs to reduce global emissions in the agricultural sector. Outlook Agric. 2014 Dec; 43(4): 241–6.
3. Baptist S, Abendroth C. A systematic review on carbon credits. 2023 [Accessed 19 May 2024]. Available from https://diglib.uibk.ac.at/download/pdf/9357953.pdf
4. Chen W, Guo Q. Assessing the effect of carbon tariffs on international trade and emission reduction of China's industrial products under the background of global climate governance. Sustainability. 2017 Jun 15; 9(6): 1028.
5. Narassimhan E, Gallagher KS, Koester S, Alejo JR. Carbon pricing in practice: a review of existing emissions trading systems. Climate Policy. 2018 Sep 14; 18(8): 967–91.
6. Kumar M, Arora N. India plans to challenge EU carbon tax at WTO [Internet]. 2023 [Accessed 2 Jul 2024]. Available from https://www.reuters.com/world/india/india-plans-challenge-eu-carbon-tax-wto-sources-2023-05-16/
7. Batra A. EU carbon tariffs are bad trade policy [Internet]. East Asia Forum. 2023 [Accessed 18 Jul 2024]. Available from https://eastasiaforum.org/2023/07/27/eu-carbon-tariffs-are-bad-trade-policy/
8. United Nations Climate Change. The Paris Agreement [Internet]. 2015 [Accessed 18 Jul 2024]. Available from https://unfccc.int/process-and-meetings/the-paris-agreement
9. Siddique AB. Impact of trade on inequality: new evidence of what, when, and where. CESifo Economic Studies. 2021 May 13; ifab005.
10. Lin W, Chen J, Gan J, Dai Y. Do firms that are disadvantaged by unilateral climate policy receive compensation? Evidence from China's energy-saving quota policy. Sustainability. 2022 Nov 18; 14(22): 15375.
11. Bundesministerium fur Wirtschaft und Klimaschutz. Germany exceeds climate financing target for emerging and developing countries [Internet]. 2023 [Accessed 1 Jul 2024]. Available from https://www.bmwk.de/Redaktion/DE/Pressemitteilungen/2023/09/20230929-deutschland-uebertrifft-klimafinanzierungsziel-fuer-schwellen-und-entwicklungslaender.html
12. Pigou A. The economics of welfare. London: Routledge; 2017.

13. Hsu SL. A complete analysis of carbon taxation: considering the revenue side. Buff L Rev. 2017; 65(4): [i]-914.
14. Steinebach Y, Fernández-i-Marín X, Aschenbrenner C. Who puts a price on carbon, why and how? A global empirical analysis of carbon pricing policies. Climate Policy. 2021 Mar 16; 21(3): 277–89.
15. Genschel P, Jachtenfuchs M. How the European Union constrains the state: Multilevel governance of taxation. European Journal of Political Research. 2011; 50(3): 293–314.
16. Government Offices of Sweden. Sweden's carbon tax [Internet]. [Accessed 2024 May 20]. Available from https://www.government.se/government-policy/swedens-carbon-tax/swedens-carbon-tax/#:~:text=Swedish%20carbon%20tax%20rates&text=The%20carbon%20tax%20was%20introduced,of%20SEK%2010.87%20per%20EUR).
17. World Bank Group. State and trends of carbon pricing dashboard [Internet]. 2024 [Accessed 19 Jul 2024]. Available from https://carbonpricingdashboard.worldbank.org/compliance/price
18. Houses of the Oireachtas. Climate | focus on the carbon tax [Internet]. 2024 [Accessed 20 May 2024]. Available from https://www.oireachtas.ie/en/press-centre/news-and-features/20240326-climate-focus-on-the-carbon-tax/#:~:text=The%20Finance%20Act%202020%20has,per%20tonne%20of%20carbon%20dioxide.
19. The effect of a carbon tax rise on Iceland's economy [Internet]. 2022 Apr [Accessed 20 May 2024]. OECD Economics Department Working Papers. Report No. 1708. Available from https://www.oecd-ilibrary.org/economics/the-effect-of-a-carbon-tax-rise-on-iceland-s-economy_76a7eaed-en
20. International Energy Agency. Carbon tax [Internet]. 2023 [Accessed 20 May 2024]. Available from https://www.iea.org/policies/16937-carbon-tax
21. Gabbatiss J. The carbon brief profile: France [Internet]. CarbonBrief. 2022 [Accessed 20 May]. Available from https://www.carbonbrief.org/the-carbon-brief-profile-france/
22. Rabbia M. Why did Argentina and Uruguay decide to pursue a carbon tax? Fiscal reforms and explicit carbon prices. Review of Policy Research. 2023 Mar; 40(2): 230–59.
23. Yoon Y. Enhancing Korea's carbon pricing mechanism [Internet]. 2024 Feb 28 [Accessed 20 May 2024]. Available from https://thedocs.worldbank.org/en/doc/c99b54dd96b03ad1fe6c2e25a37c149c-0320072024/original/Session-3-1-Yeochang-Yoon-KDI-F.pdf#page=1.00
24. Allayannis G, Tenguria SK. Carbon credit markets. SSRN Journal [Internet]. 2009 [Accessed 19 May 2024]. Available from https://www.ssrn.com/abstract=1419269
25. Sato M, Rafaty R, Calel R, Grubb M. Allocation, allocation, allocation! The political economy of the development of the European Union Emissions Trading System. WIREs Climate Change. 2022 Sep; 13(5): e796.
26. European Parliament and Council of the European Union. Directive 2003/87/EC of the European Parliament and of the Council of 13 October 2003 establishing a scheme for greenhouse gas emission allowance trading within the Community and amending Council Directive 96/61/EC (Text with EEA relevance) [Internet]. 2003 [Accessed 19 May 2024]. Available from https://eur-lex.europa.eu/eli/dir/2003/87/oj

27. Paterson M. Who and what are carbon markets for? Politics and the development of climate policy. Climate Policy. 2012; 12(1): 82–97.
28. Anderson-Teixeira KJ, Belair EP. Effective forest-based climate change mitigation requires our best science. Global Change Biology. 2022 Feb; 28(4): 1200–3.
29. Badgley G, Freeman J, Hamman JJ, Haya B, Trugman AT, Anderegg WRL, et al. Systematic over-crediting in California's forest carbon offsets program. Global Change Biology. 2022 Feb; 28(4): 1433–45.
30. Cannon J. Forest carbon credits and the voluntary market: a solution or a distraction? [Internet]. MONGABAY. 2024 [Accessed 19 May 2024]. Available from https://news.mongabay.com/2024/01/forest-carbon-credits-and-the-voluntary-market-a-solution-or-a-distraction/
31. West TAP, Börner J, Sills EO, Kontoleon A. Overstated carbon emission reductions from voluntary REDD+ projects in the Brazilian Amazon. Proc Natl Acad Sci USA. 2020 Sep 29; 117(39): 24188–94.
32. Takahata K, Suetsugu H, Fukaya K, Shirota S. Bayesian state-space synthetic control method for deforestation baseline estimation for forest carbon credits. Environ Data Science. 2024; 3: e6.
33. Summers SK, Rainey R, Kaur M, Graham JP. CO_2 and H_2O: understanding different stakeholder perspectives on the use of carbon credits to finance household water treatment projects. PLoS ONE. 2015 Apr 30; 10(4): e0122894.
34. 'Get out of jail free': how plastics offsetting is giving industry a license to pollute. [Internet]. 2023 [Accessed 20 May 2024]. Available from https://www.source-material.org/plastic-offsetting-philippines-pcx-verra-cement/
35. Garsia A, Moinet A, Vazquez C, Creamer RE, Moinet GYK. The challenge of selecting an appropriate soil organic carbon simulation model: a comprehensive global review and validation assessment. Global Change Biology. 2023 Oct; 29(20): 5760–74.
36. Köhl M, Baldauf T, Plugge D, Krug J. Reduced emissions from deforestation and forest degradation (REDD): a climate change mitigation strategy on a critical track. Carbon Balance Manage. 2009 Dec; 4(1): 10.
37. Mandatory new requirements for sustainable construction [Internet]. 2023 [Accessed 23 May 2024]. Available from https://global.eg.dk/news/2023/april/mandatory-new-requirements-for-sustainable-construction/
38. Fabris P. France to mandate all new public buildings be 50% timber or other natural materials [Internet]. 2022 [Accessed 31 Mar 2024]. Available from https://www.bdcnetwork.com/france-mandate-all-new-public-buildings-be-50-timber-or-other-natural-materials
39. Government of Canada. Greening government strategy: a government of Canada directive [Internet]. Available from https://www.canada.ca/en/treasury-board-secretariat/services/innovation/greening-government/strategy.html
40. City of Helsink/Urban Environment Division. The city of Helsinki's roadmap for circular and sharing economy [Internet]. 2020 [Accessed 4 Jul 2024]. Available from https://circulareconomy.europa.eu/platform/sites/default/files/the-city-of-helsinkis-roadmap-for-circular-and-sharing-economy.pdf
41. City of Helsinki. The city of Helsinki's roadmap for circular and sharing economy [Internet]. [Accessed 23 May 2024]. Available from https://circulareconomy.europa.

eu/platform/sites/default/files/the-city-of-helsinkis-roadmap-for-circular-and-sharing-economy.pdf#page=14.10

42. Votre gouvernment. Policy for the use of wood in construction [Internet]. [Accessed 4 Jul 2024]. Available from https://cdn-contenu.quebec.ca/cdn-contenu/forets/documents/entreprises/PO_wood_construction_MRNF.pdf
43. Douro_Dummer. Sustainable Development Program [Internet]. [Accessed 23 May 2024]. Available from https://www.dourodummer.ca/en/building-and-renovating/sustainable-development-program.aspx
44. 2000 Watt Site. 2000-watt society and 2000-watt site [Internet]. [Accessed 29 May 2024]. Available from https://www.2000watt.swiss/english.html
45. Luo D, Tang D, Zhou B. Decarbonising the built environment in China Integrating the efforts of regulators and companies [Internet]. London: Royal Institution of Chartered Surveyors (RICS); 2023 Nov [Accessed 25 May 2024]. Available from https://www.rics.org/content/dam/ricsglobal/documents/reports/rics-cop28-policy-paper-china.pdf
46. Gurvinder Pal Singh R. Decarbonising the built environment in India—addressing operational and embodied carbon. London: Royal Institution of Chartered Surveyors; 11/20203.
47. Bureau of Indian Standards. CED 2 New Standard List [Internet]. 2023 [Accessed 30 May 2024]. Available from https://www.services.bis.gov.in/php/BIS_2.0/dgdashboard/Published_Standards_new/standards?commttid=MTkw&commttname=Q0VEIDI%3D&aspect=&doe=&from=2022-07-03&to=2023-07-03
48. Establishment of Indian Database of Embodied Energy/Carbon in Building Construction Materials [Internet]. [Accessed 30 May 2024]. Available from https://www.carbse.org/research_development_view?page=17

Chapter 8

Low-tech and low-process approaches to low-embodied carbon buildings

Learning from traditional, vernacular, and indigenous knowledge

8.1 Principles of low-tech and low-process approaches

Low-tech and low-process approaches to reducing buildings' embodied carbon encourage us to rethink sustainability in built environment. By focusing on minimizing the carbon footprint associated with materials and processes, we can achieve significant gains without heavily relying on cutting-edge technologies. Valuable lessons from the Global South, where resource-efficient and sustainable building practices have been refined out of necessity over centuries, can guide us towards a greener future. Embracing these time-tested methods honours traditional wisdom and paves the way for innovative, sustainable solutions in modern construction.

The first crucial principle we can learn from the Global South is *designing a system* rather than the sum of its parts, which involves reducing complexity and optimizing functionality. This involves creating an integrated system where every building component works synergistically to improve overall performance and sustainability. By reducing complexity and optimizing functionality, we enhance both the cultural and aesthetic presentation of buildings while minimizing the resources required for construction and maintenance.

The second important principle is *designing for longevity*. Extending the usable life of buildings and their components is a powerful way to reduce embodied carbon. Wiik et al. shows that increasing the lifespan of buildings can slash embodied carbon by up to 20% [1]. This is achieved through durable materials, adaptable design features, and construction techniques that make maintenance and upgrades accessible. Examples from the Global South demonstrate how robust designs and local materials lead to structures that are both enduring and adaptable.

Embodied Carbon. Ming Hu, Oxford University Press. © Ming Hu (2025). DOI: 10.1093/9780198944485.003.0008

The third principle involves *utilizing localized, low-cost, and low-carbon solutions*. The innovative use of local materials in the Global South often goes unnoticed in modern construction practices, yet these materials offer a treasure trove of sustainable solutions. This oversight can be attributed to a Global North bias in knowledge generation, which will be explained in detail in Section 8.1.3. By embracing the principles and knowledge from the Global South, we can create sustainable and cost-effective buildings.

In essence, learning from the Global South can enhance our understanding and application of these principles, offering tried and tested methods for achieving sustainability. In the following sections, we will explore these principles in greater detail and delve into two case studies that showcase their practical application.

8.1.1 Design for a system

Traditional, vernacular, and indigenous designs aim to be more than the sum of their parts, embodying practices and principles deeply rooted in the culture of their communities. As Amos Rapoport suggests in *House Form and Culture*, house form is 'a physical mechanism which reflects and helps to create the worldview, ethos, and so on, of a people, comparable to the various social institutions (or mechanisms) which do the same' [2]. Unlike mainstream modern Global North architecture, which often evaluates building components separately according to certain criteria from different disciplines, such as structure, fire protection, energy performance, architecture style, and others. Moreover, cultural relevance and appropriateness as an integral part of traditional design are not measured and included in any current green building rating system.

Traditional designs often leverage readily available resources, efficiently using local materials while creating cultural identity. For example, the tipi, a traditional structure used by nomadic tribes of the Great Plains in North America, epitomizes this systematic approach, where structure, comfort, adaptability, and cultural relevance are considered as a whole. Let us first take a look at the tipi's construction. As illustrated in Figure 8.1, the tipi, cone-shaped and averaging 14 to 16 feet in diameter and about 17 feet tall, was typically constructed with twenty wooden poles covered with twelve to fourteen fabricated buffalo hides, which were once plentiful on the Plains [3]. The simplicity and effectiveness of this design supported the nomadic lifestyle, enabling quick assembly and disassembly.

Figure 8.1 Traditional tipi. Tipi from outside view.

When we look at the tipi from the inside, a notable refinement in the tipi's design is the dew cloth—an interior lining hung by ropes tied to the middle of the tipi poles. Its design allowed for seasonal adjustments, such as sod and snow packing in winter and elevated ventilation in summer. This feature protected inhabitants from harsh weather by promoting ventilation and creating insulation, keeping the tipi warmer in winter and cooler in summer, making the structure highly adaptable to the nomadic lifestyle.

Besides the integrated function, form, and physical performance, cultural relevance is central to indigenous architecture, such as a tipi, which embodies practices and principles informed by the community's respect for the relationship between man and nature. Indigenous motifs are not random or merely decorative; they relate to cultural symbols integral to representing and depicting the worldview of indigenous people. For example, in the design of tipis, entrances oriented to the east symbolize spiritual renewal, paying homage to the rising sun as a source of renewal and birth. The Blackfoot tribe's tipis, primarily constructed by women, served both utilitarian and spiritual purposes.

The emphasis on circular movement within the tipi was efficient and reflected the strong spiritual value placed on the circular form. The circle, considered a sanctified movement pattern with no beginning and no end, is central to Native American beliefs. Edward Curtis interprets this belief: 'everything the Power of the World does is done in a circle . . . The life of a man is a circle from childhood to childhood, and so it is in everything where power moves'. Adolf Hungrywolf, a Germany-born writer who is known for his dedication to preserving Native American history and tradition, has provided evidence that the circular shape of indoor space stimulates a specific experience for dwellers. One modern tipi dweller confirmed that 'the roundness of our dwelling gives me a very strong feeling—like being inside a geometric molecule and tumbling through the vastness of the Universe—physical old age next to me, spiritual, everlasting age all around' [4].

In addition to providing a unique interior spatial experience, the tipi facilitated a connection between the dwelling and the surrounding environment, the open plain. The light filtering through the semi-translucent cover allowed inhabitants to follow the sun's progression, while the smoky fire and incense from sacred trees recreated forest scents. Tipis were often decorated with paint, symbolizing the connection with nature. These embellishments were not merely decorative but indicated a particular tipi as a convergence point for potent spiritual power.

What is particularly interesting and inspiring is that such a systematic and integrated approach resulted in something that can be defined as sustainable. Here, the definition of sustainability used is from the United Nations World Commission on Environment and Development in 1987: 'Development that meets the needs of the present without compromising the ability of future generations to meet their own needs' [5]. According to the U.S. Environmental Protection Agency, 'Native Americans were the first green architects and builders of the Americas, and traditional Native American building designs and practices are sustainable' [6]. The holistic design principles in traditional, vernacular, and indigenous architecture reflect a deep understanding of local materials, functionality, flexibility, spiritual experience, and cultural relevance. These designs create practical and comfortable living spaces while maintaining a strong connection to cultural heritage and the natural environment. This integrated approach contrasts with the segmented methods often seen in modern Global North architecture. Studying and appreciating these designs teaches valuable lessons in sustainability and resource efficiency applicable to modern construction practices.

8.1.2 Design for longevity

Designing buildings to last involves considering the entire life cycle from the onset to ensure durability and long-term performance. The common perception in modern building practices, dominated by the Global North approach, is to rely on durable materials like reinforced concrete and corrosion-resistant metals, which are resistant to wear and tear, weather conditions, and potential environmental stresses. However, these materials are often embodied carbon-intensive and expensive [7]. This raises the question: if relying solely on these modern materials is not viable, what other options do we have?

We should explore alternative building materials that are inspired by traditional and vernacular architecture in the Global South, such as bamboo, which is durable, renewable, and affordable [8]. However, incorporating these 'Global South' materials requires not only adjustments to building codes but also a shift in the Global North-dominated perspective on construction methods. To a large extent, acknowledging that the Global South can lead the way to sustainability is most challenging to the Global North dominance mindset, consequently creating blind spots, sometimes even hindering the real process of moving towards carbon emission reduction. Let's take wood as an exemplary example. Wood, known for its low embodied carbon, is a key strategy in the Global North for reducing embodied carbon, and is often encouraged as a replacement for concrete and steel. However, sustainable wood, particularly from responsibly managed forests, is less common in many regions of the Global South due to geographic limitations and other factors. As illustrated in Figure 8.2, the largest forests supporting sustainable wood sourcing are predominantly in the Global North, specifically North America and Europe, accounting for 74% of the supply. In contrast, Asia has only 3% supply, Latin America 11%, and Africa 6%. Therefore, it is not surprising that some Global South countries are ahead in adopting other low-embodied carbon materials, like bamboo. For example, China and India included bamboo as a major structural material in their building code, a practice not yet embraced by Global North countries [9]. Later in Section 8.3.2, a case project will be used to demonstrate the application of bamboo as a major construction material.

However, using low-embodied carbon building materials only addresses part of the problem. Extending buildings' useful lifespans can significantly reduce carbon emissions by reducing the demand for new construction. This

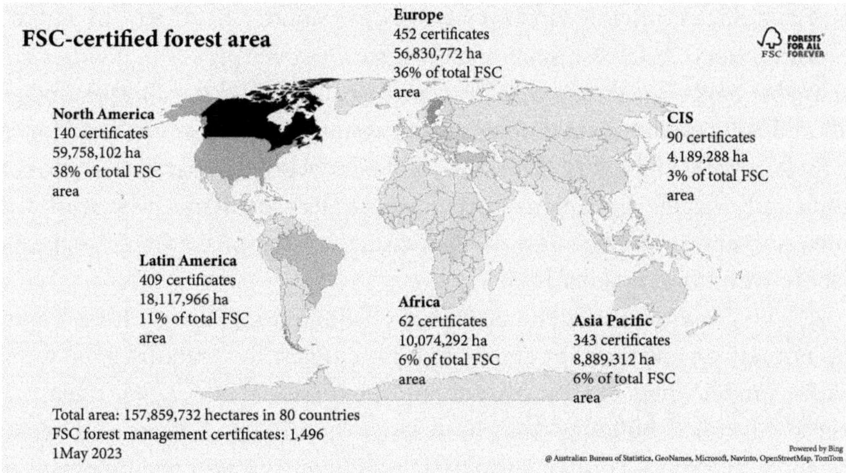

Figure 8.2 Reproduced from Forest Stewardship Council UK, 'FSC-Certified Forests by Region' under a Creative Commons Attribution 4.0 International https://uk.fsc.org/what-is-fsc/facts-and-figures (CC BY 4.0).

requires that, from the beginning, the design of buildings with maintenance in mind. This involves ensuring easy access to critical systems for inspection and repairs, using modular components that can be easily replaced, and providing detailed maintenance manuals. Additionally, designing buildings to be adaptable to changing needs makes them more likely to remain useful over time. Flexible floor plans, multifunctional spaces, and easily modifiable structures allow buildings to evolve with their occupants' needs, thereby extending their usability.

Construction quality is equally important, playing a critical role in the longevity of a building. Skilled labour, attention to detail, and adherence to best practices ensure that buildings are constructed correctly from the start, reducing premature failures and embodied carbon derived from the repair needs. However, modern society prioritizes efficiency and speed, especially in construction and manufacturing, where mass production is encouraged for economic gain. This focus often overlooks craftsmanship and low-tech methods. Historically, buildings constructed before World War II using traditional methods and materials, such as unreinforced load-bearing masonry, are regarded as high-quality and well-built, requiring minimal repairs [10]. In contrast, post-World War II construction, which increased rapidly to support economic recovery and population growth, often varies in quality and durability compared to pre-war structures [11]. The emphasis on mass production and cost efficiency has led to lower quality, exemplified by American post-war housing, where brick walls are often just facades over plywood, giving a false sense of quality and strength. Contractors frequently use the cheapest materials and build quickly to maximize profits, resulting in subpar construction. As a *Forbes* article highlights, 'Workmanship has declined in parallel. There continue to be expert craftsmen—carpenters, roofers, painters—who work with precision and pride, but they are increasingly being pushed out by cheaper labour with inferior skills' [12].

All the design considerations, craftsmanship, and skills mentioned above do not rely on advanced or emerging technologies; they require good design skills, fundamental quality control, and professional ethics. We often marvel at historical buildings that have lasted hundreds of years and remain in good condition. Despite basic design and construction technology at the time, their longevity was ensured by careful design and planning, high-quality craftsmanship, professional ethics, and consistent maintenance. Extending the lifespan of the existing building and reducing the newly built area is a direct and effective strategy, as it avoids carbon emissions across all life-cycle stages. This can be achieved by creating multifunctional spaces and efficiently using existing ones. Efficient use of the existing built stock can decouple building demand growth from economic development, making it a viable consumption-oriented strategy [13].

8.1.3 Localized, low-cost, and low-carbon materials

As mentioned earlier, wood is not readily available as a building material in many locations, especially in the Global South, due to forest area distribution. The developed regions in the Global North have much lower rates of new construction and demand for building materials [14]. In contrast, the global building stock is expected to double, with significant increases in Global South developing countries in Africa and Asia, driven by substantial population growth [14], where wood as a construction material is often inaccessible. By 2055, Asia's population is projected to peak at 5.4 billion, while Africa's population is expected to reach 10.9 billion by 2100 [15]. In these rapidly growing regions, the demand for wood as a primary construction material cannot be met. Moreover, the reliance on high-tech and high-cost solutions like cross-laminated timber is not accessible to most developing countries in the Global South. Despite the impracticality of mass timber construction in many regions, research and development of this material continue to be heavily promoted and funded by numerous Global North agencies.

If we are serious about finding accessible, practical, and equitable solutions for Global South countries, we need to shift our focus from a Global North perspective to an inclusive approach that draws knowledge from traditional and vernacular architecture in the Global South. Vernacular architecture, defined as 'the architectural language of the people with its ethnic, regional, and local dialects: the product of non-experts', responds to local climatic, material, and craft conditions while reflecting the local community's culture, customs, and lifestyle [16]. A key characteristic of vernacular buildings is their use of bio-based, low-embodied carbon materials, such as bamboo, straw, and soil [17]. These materials convert CO_2 into biomass through photosynthesis during the plant's growth before being processed into building materials [18]. Soil carbon sequestration, where CO_2 is transferred from the atmosphere and stored in the soil through plants, is another important aspect of using bio-based materials [19]. Consequently, traditional earth buildings in the Global South are inherently low in embodied carbon. As illustrated in Figure 8.3, mud brick (e.g. adobe brick) has lower embodied carbon than timber (i.e. wood), straw has the lowest embodied carbon, and natural stone has much lower embodied carbon than other traditional and contemporary materials. By embracing vernacular architecture and bio-based materials, we can develop more sustainable, culturally relevant construction practices suited to the Global South's needs and resources.

We can divide the low-embodied carbon materials inspired by traditional and vernacular buildings in the Global South into three categories: primary natural materials, secondary natural materials, and low-processed materials [20]. *Primary natural materials* are those found in the natural environment and applied directly to building construction. Examples include wood, stone, bamboo, mud brick (natural dry), and grass. These materials are popular across various climatic regions and are used globally due to their avand embracing more sustainableailability and low cost. *Secondary natural materials* involve processing natural materials from other industries to create building materials or components. For example, waste wood from logging and agricultural residues like straw are repurposed for construction. Many materials from both the primary and secondary categories sequester CO_2 during their growth, giving them a negative carbon footprint, which is beneficial for reducing overall carbon emissions. The third category includes *low-processed materials* like fired brick, terracotta tiles, and burnt wood. These materials require minimal processing, such as kiln drying. In the following, we will explore one exemplary material from each category; for a longer and more detailed explanation, readers can consult the author's article entitled 'Exploring Low-Carbon Design and Construction Techniques: lessons from vernacular architecture' [20].

Primary natural materials: adobe

Adobe blocks are among the earliest and most commonly used natural building materials worldwide. They are utilized in regions such as Asia [21], Europe [22], North America [23], the Middle East, and Africa [24] due to the widespread availability of their raw ingredients: clay, sand, and water (Figure 8.4, left). Various natural and industrial materials, such as straw and gravel, can be added to enhance the properties of adobe blocks, improving moisture resistance, durability, and overall construction quality while reducing their environmental impact. Adobe blocks are sun-dried and used to build thick masonry walls [25]. Although straw is added in small amounts, it plays a crucial role in binding the adobe blocks and ensuring even drying to prevent cracking from uneven shrinkage [26], because when straw is combined with earth and kept moist for a period of two to three days, it produces lactic acid. This lactic acid acts as a natural stabilizing agent, which helps to strengthen or stabilize the mixture [27].

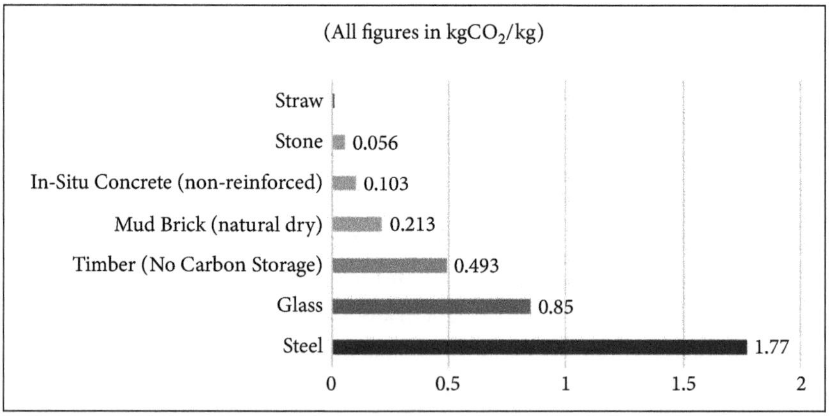

Figure 8.3 Embodied carbon of building materials.

Source: date from Inventory of Carbon Energy (ICE) https://circularecology.com/embodied-carbon-footprint-database.html

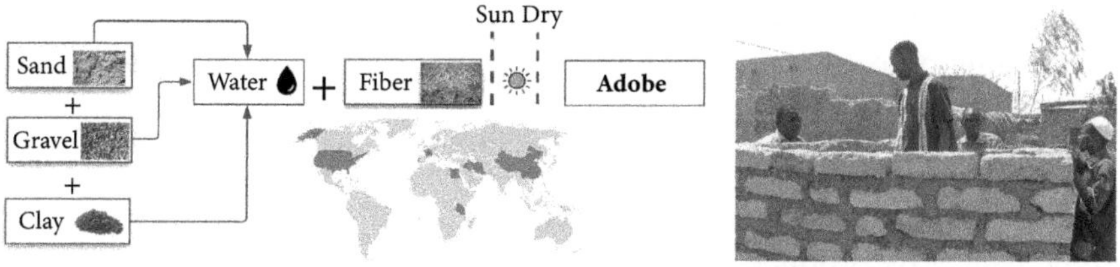

Figure 8.4 Adobe block (left) ingredients and making process. (Right) Building an adobe block wall.

Adobe is low cost, locally available, recyclable, and adaptable to various soil types [20]. It also offers good thermal and acoustic properties, making it a practical choice for many builders [28]. In 1964, the Ministry of Science and Research in Cairo invited architects and engineers to build a model room on the grounds of the Research Centre in Cairo. Two extreme cases were constructed—one in mud brick (adobe) and the other in prefabricated concrete slabs. When evaluated from a climatic control perspective, it was found that the air temperature in the prefabricated concrete room was 11°C (52°F) higher than in the mud-brick room, and the mud-brick house remained within the comfort zone without any mechanical air conditioning [27]. In addition, the ease of construction with adobe reduces labour, equipment requirements, energy consumption, and emissions, making it accessible to many people (Figure 8.4, right).

Despite these advantages, adobe does have disadvantages, such as low seismic strength [23]. However, the primary reason for the decline in the use of adobe blocks in the modern era is not structural deficiency but social and psychological perceptions. Adobe is often viewed as a 'poor man's material' or primitive, especially in the contemporary Global North perspective. This perception is far from the truth, as adobe has been successfully used in Global North countries until the modern era with the introduction of embodied-intensive materials like concrete and cement. For example, in Portugal, adobe was widely used in almost all types of construction until the mid-twentieth century when the cement industry developed [22]. In recent years, research has been conducted on adobe, but the focus has primarily been on building conservation and historical preservation rather than treating adobe as a viable building material again. This shift in focus overlooks the potential of adobe as a sustainable, cost-effective, and environmentally friendly building option for modern construction. By reconsidering and promoting the use of adobe, we can leverage its many benefits while challenging outdated perceptions and embracing more sustainable building practices.

Secondary natural materials: straw bale

Straw bale, a by-product of harvested plants, has found a surprising second life in the world of construction. Composed of dry plant stalks left behind after crops like wheat, oats, barley, rye, rice, soybeans, and corn have been harvested [29], straw bale construction boasts a rich history. Remarkably, the use of straw in buildings dates back to the Paleolithic Period in Africa [29, 30]. Its utility has spanned continents and centuries, becoming a staple in the vernacular architecture of the Middle East and China for thousands of years [31]. Early settlers in treeless regions or areas with unsuitable soil for mud houses found straw to be the best and often the only practical building material. Straw bales can be used exposed or covered with soil or lime stucco, depending on the needs of the construction. The shape, dimension, and compression level of straw bales are determined by the type of baler used [32]. Research, such as those conducted by Goodhew et al. and Margues et al. have shown that the thermal conductivity of straw bales is lower than that of concrete, brick, and wood, making them an excellent insulator. Additionally, their specific heat capacity ranges between 1,075 and 2,000 J/(kg·K), comparable to conventional materials like concrete and brick, which range between 1,000 and 2,500 J/(kg·K) [33, 34].

In traditional vernacular architecture, straw bales were often used as load-bearing wall materials, mixed with soil and other earthy materials, showcasing their remarkable versatility and strength. A classic example is the 'Nebraska-style' straw bale wall in North America, which provided enough structural integrity to support the weight of the roof while also offering excellent insulation. Multiple studies have highlighted straw bales' outstanding qualities for building high-performance exterior walls. Its superior thermal properties and good hygrothermal performance make it an ideal material for sustainable construction [32, 34].

Straw bales have taken on a different role in modern construction in the Global North. Figure 8.5 illustrates that they are more commonly used as insulative infill than load-bearing elements. The straw bale core is typically surrounded by outer layers made of hardwood, plaster, or cement, enhancing the structure's durability and aesthetic appeal [32]. Overall, straw bale construction is cost-effective, environmentally friendly, and a charming blend of ancient wisdom and modern efficiency.

Figure 8.5 Straw bale wall.

Reproduced from Colin Rose via Wikimedia Commons under a Creative Commons Attribution license (CC BY 2.0).

Low-processed materials

Compared to traditional mud brick, the modern method of brick manufacturing is notably carbon-intensive due to the high-temperature firing process and the decomposition of carbonates [35, 36]. Weyant et al. found that the brick industry annually emits substantial amounts of carbon monoxide (CO), particulate matter (PM2.5), and other pollutants, underscoring its significant environmental impact [35]. The large-scale CO_2 emissions from fired-brick production further highlight the urgency of addressing these environmental concerns [37]. Additionally, biomass co-fired brick kilns contribute to elevated concentrations of health-relevant non-methane volatile organic compounds (NMVOCs) like benzene [38], exacerbating the environmental and health implications of conventional brick production. These high emissions of particulate matter and other pollutants are primarily due to the type of fuel used in the firing process, contributing to air pollution and associated health risks [39].

To produce low-process fired bricks, the key is to reduce the firing temperature and time, consequently reducing energy use and carbon emission. One promising approach, inspired by indigenous building traditions and knowledge, involves incorporating organic waste or sewage sludge into fired-clay bricks, which helps reduce their carbon footprint [40, 41]. Studies have shown that using treated wastewater sludge (biosolids) in fired-clay bricks can decrease energy demands during firing, presenting a more environmentally friendly manufacturing method [42, 43]. Various waste materials, such as organic waste, waste treatment sludge, fly ash, cigarette butts, rice husk, and processed waste tea, have been successfully integrated into fired-clay bricks, demonstrating the potential for eco-friendlier building materials [44, 45].

For example, in India and Sri Lanka, Rodney et al. and Kulshreshtha et al. discovered that adding cow dung ash to the brick mixture improves the durability and moisture resistance while reducing firing time [46, 47]. Furthermore, incorporating additives like wood ash or palm oil waste into fired-clay bricks has been found to shorten the firing time and improve specific properties such as water absorption and compressive strength [44]. These innovations demonstrate the potential for more sustainable brick production methods, contributing to reduced environmental impact and improved building material performance.

8.2 Design strategies for low embodied carbon

8.2.1 Embracing sustainable building through modular design

The Construction Industry Institute defines modular construction as 'the use of offsite construction (including a segregated area onsite) and includes all work that represents substantial offsite construction and assembly of components and areas of the finished project' [48]. Replacing on-site construction with prefabricated modular components has a long tradition and has proven to be an effective method for reducing carbon emissions [20]. This approach, often seen as a modern innovation, minimizes construction waste and boosts efficiency. Yet, this technique has roots stretching back thousands of years, embedded in traditional and vernacular architecture across various cultures and periods.

Imagine ancient Roman cities bustling with activity. Among the many architectural marvels were shops, warehouses, cisterns, and military structures built

Figure 8.6 (A) Surface remains of the southern part of building S3; view from western side. (B) Trench W1-5 with remains of Building W1-B; view from western side. (C) Building W2; view from eastern side. (D) Western part of double bath complex T2; view from northeast.

Reproduced from Gwiazda, M. 'Modular Designs at the Early Byzantine Pilgrimage Site of Philoxenite, Egypt' (2023), *Journal of Roman Archaeology*, doi: https://doi.org/10.1017/S1047759423000120. (Courtesy of the Polish Center of Mediterranean Archaeology; photos by M. Gwiazda.) Under a Creative Commons Attribution license 4.0 (CC BY 4.0).

using modular design. Architects created buildings composed of replicated segments, making it easier to visualize and estimate capacities. In the port of Ravenna, for example, eighteen identical warehouses were constructed in the early fifth century CE, showcasing the practicality and efficiency of modular construction [49]. Similarly, as illustrated in Figure 8.6, the Monastery of Abba Nefer in Middle Egypt, dating back to the fifth–sixth century CE, featured rooms with similar layouts, potentially used as monks' cells. This method was not just about practicality but also about ensuring uniformity and functionality across various structures [49].

Now, let's travel to the sixth century CE, to the town of Philoxenite in northern Egypt. This settlement, built to serve pilgrims traveling to the shrine of Saint Menas in Abū Mīnā, stands as a testament to the large-scale application of modular design. Gwiazda and colleagues revealed six large building complexes based on modular principles, designed to meet the needs of the pilgrims. This approach not only facilitated urban planning but also allowed for a diverse range of building plans to be replicated across the town. It was a brilliant solution to the challenges of large-scale construction and urban planning [49].

But modular design wasn't confined to civilian structures. It played a crucial role in military architecture too. Take the early second-century Roman auxiliary fort at Humayma, for instance. Excavations revealed the use of standardized measurements, suggesting a highly organized approach to planning and construction. By adhering to consistent units of measurement, Roman engineers could quickly and effectively establish fortifications, which were vital for maintaining control over vast territories [50].

Across the globe, in China during the Song Dynasty, the publication of the Ying Zao Fa Shi ('construction method') in 1103 CE marked a significant advancement in modular construction [20]. This manual defined wood modules based on size, which were then used to construct buildings with various functions and hierarchies. For example, as illustrated in Figure 8.7 (lower image), the first-grade wood module was used for large temples, while smaller modules (Grade 8) were suitable for utility buildings. These individual modules were made off-site and transported to the construction site for assembly, much like modern prefabricated houses. Yet, such a high modularity approach yields cultural richness in traditional Chinese architecture, as demonstrated in Figure 8.7 (upper image).

Another captivating example of modular and sustainable architecture is found in the bamboo houses of Southeast Asia. In the Chittagong Hill Tracts of Bangladesh, bamboo is the primary material for floors, columns, and exterior walls (Figure 8.8) [51]. The local building practices here are ingenious, with the lashing method allowing for remarkable flexibility. These structures can be easily disassembled, expanded with additional bamboo, or entirely rebuilt in a short time. The bamboo houses, demonstrate a logical approach to construction, highlighting the systematic and efficient use of local materials to create adaptable and sustainable living spaces [52, 53].

The historical adoption of modular design across different regions and cultures underscores its significance in facilitating large-scale construction and urban planning in response to specific needs. Modular design and construction are inherently sustainable and deeply rooted in traditional architecture. By relearning and adapting these ancient techniques, we can enhance modern building practices, promote efficiency, reduce waste, and contribute to environmental sustainability.

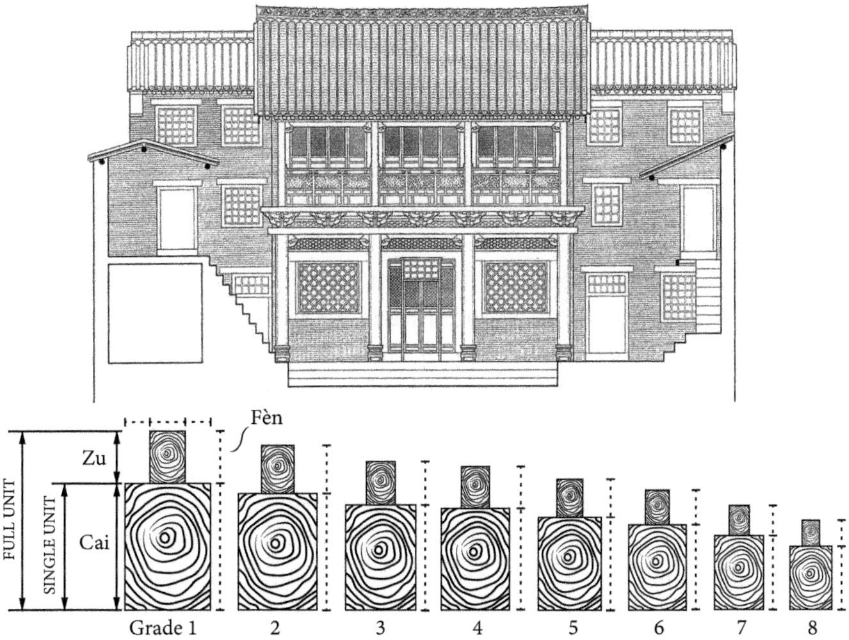

Figure 8.7 Wood modules in traditional Chinese buildings.

Figure 8.8 Bamboo construction in Chittagong Hill Tracts, Bangladesh.

8.2.2 Designing for disassembly and reuse

The United Nations defines a circular economy as a system where products' lifespans are extended through better design and where waste is redirected from the end to the beginning of the supply chain [54]. In the building and construction industry, one powerful way to embrace this concept is by reusing components from old structures into new ones and using more durable and recycled materials. This can be effectively achieved through the concept of design for disassembly. Designing buildings with disassembly in mind can lead to significant environmental benefits. In some instances, this approach has been linked to reductions of up to 81% in embodied energy and 88% in embodied carbon [55]. However, despite these advantages, the practice of designing for disassembly is not yet widespread in the construction industry. Two main challenges hinder its adoption in the modern Global North construction sector: the inflexibility of modern buildings and the complexity of disassembling current construction materials.

Most modern buildings are designed for a single purpose, lacking the flexibility and adaptability required for disassembly and reuse. Additionally, the construction techniques and materials used today make disassembling it extremely difficult. For example, recycling concrete requires separating it from steel rebar through crushing and magnetic separation or mechanical removal, which makes the process labour-intensive, costly, and cumbersome [56]. Consequently, nearly all reinforced concrete ends up demolished rather than disassembled at the end of its life cycle.

Traditional and vernacular architecture offer valuable lessons in this regard. Historically, design for disassembly was a common practice in traditional timber construction worldwide. In medieval Europe, the scarcity of suitable timber necessitated the regular reuse of major structural members from one building to another. Similar practices were observed in Asia, where traditional Chinese and Japanese wooden houses were constructed with primary and secondary frames that could be easily disassembled and reassembled using simple tools, as they typically did not use nails or screws. The Japanese term 'kaitai shūri', meaning 'repair by disassembly', reflects a tradition where buildings could be partially or entirely disassembled and reassembled, preserving both primary and secondary structures. Before 1897, kaitai shūri was performed by carpenters, but since then, academically trained architects and engineers have taken over this practice from craftspeople [57].

In the early twentieth century, Japan developed an interest in surveying disassembled parts of buildings to learn more about the technological characteristics of different periods. This newfound knowledge was used to classify Japanese shrines and temples and to document the findings of academic surveys conducted by architectural scholars. This tradition of disassembling and reassembling buildings has endured throughout history and can be applied to various building types. For example, although there are no well-documented studies or practices, it is possible to disassemble adobe blocks from plaster-free walls for reuse in other buildings. These historical practices demonstrate the potential of design for disassembly and highlight the need for modern architects to integrate these principles into contemporary design.

8.3 Two case studies

Embracing the age-old techniques mentioned in section 8.2 can pave the way for a more sustainable and efficient built environment where resources are conserved and environmental impacts are minimized. To illustrate how traditional knowledge and techniques from the Global South can be used to build sustainable buildings appropriate for Global South conditions without relying on Global North techniques and materials, two exemplary buildings are used to demonstrate how and why this is smarter and truly sustainable.

8.3.1 Blend of tradition and innovation: Gando Primary School

In the heart of Burkina Faso lies Gando Primary School, designed by the architect Diébédo Francis Kéré and exemplifying the harmonious integration of traditional construction techniques with sustainability goals. Kéré, originally from Gando, has a deep understanding of local needs and resources [58]. His design for Gando Primary School utilizes locally sourced clay and passive cooling strategies inspired by indigenous knowledge. This approach not only reduces the need for complex mechanical systems but also creates a building that is both functional and sustainable.

The primary goal of the school's construction was to expand the limited educational infrastructure in Boulgou province. Local schools often suffer from poor lighting and ventilation, issues that Kéré adeptly addressed within the constraints of cost, climate, resource availability, and construction feasibility. Clay, abundant in the region and traditionally used in house building, was

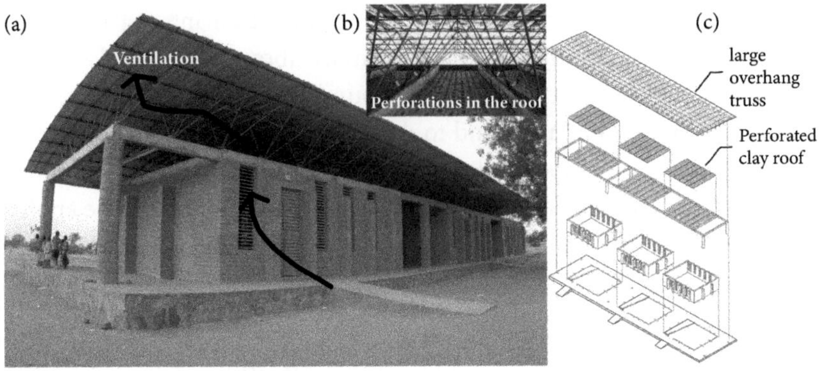

Figure 8.9 (a) Building perspective with ventilation diagram. (b) Perforated clay brick roof. (c) Modular system.

combined with 10% cement to create robust bricks capable of withstanding heavy rain. These bricks provided thermal protection against the hot climate and were easy to produce locally.

To protect the walls from heavy rain, a large overhanging roof was designed, as shown in Figure 8.9, which gives this building a distinguished character. In Burkina Faso, corrugated metal roofs are common in modern buildings due to their ability to shed heavy rainfall during the wet season. However, they make interiors extremely hot during the dry season. Kéré resolved this by separating the roof from the interior learning spaces with a dry-stacked brick ceiling, allowing for maximum ventilation. As illustrated in Figure 8.9a, cool air is drawn in through windows, while hot air escapes through perforations in the clay roof (Figure 8.9b), creating a comfortable environment without a mechanical system. As Kéré stated, '[We created] a well-ventilated structure, which is working without any artificial or mechanic climate control' [59]. This design significantly reduces the school's ecological footprint by eliminating the need for air-conditioning, which is often deemed as necessary from the modern Global North perspective.

After finding out the design solution of building a large overhanging roof to protect brick walls and improve ventilation, the next challenge is to construct it. Transporting large elements to the site was impractical due to cost and infrastructure limitations. Therefore, Kéré designed a basic modular construction system using simple steel bars to create lightweight trusses, which were then topped with metal sheeting for the roof, as demonstrated in Figure 8.9c. This design choice facilitated construction by local villagers, who were trained to use simple tools such as handsaws and welding machines. This not only introduced new low-tech building technologies inspired by traditional knowledge but also provided young members with construction skills they have since used in other projects. This project not only stands as a beacon of sustainable architecture but also as a symbol of what can be achieved when traditional knowledge is integrated with innovative application, providing a blueprint for the future of sustainable construction in the Global South.

8.3.2 Community built with bamboo: the story of Green Village

Nestled in the lush landscapes of Bali, Indonesia, lies Green Village, a pioneering living community developed by John and Cynthia Hardy. This innovative project features twelve bamboo houses, designed by the IBUKU architectural team led by their daughter, Elora Hardy. The Hardys also initiated the Green School project, which was made entirely from bamboo. They discovered this versatile material after moving from Canada to Indonesia; without formal training as architects or engineers, they learned to design and build bamboo houses alongside the local community.

In the Global North, particularly in North America, bamboo is often seen as a harmful invasive plant. For instance, in March 2022, the Fairfax County (Virginia, U.S.) Board of Supervisors passed an ordinance requiring bamboo owners to prevent its spread to public areas or neighbouring properties [60]. From this perspective, bamboo is not considered a sustainable or useful material; its presence is often contested and eradicated. Conversely, in the Global South, bamboo is viewed quite differently. As Defit Wijaya, IBUKU's chief architect, explains, 'We used bamboo simply because we have it in Indonesia . . . it is part of our cultural heritage, and we have been using it for a long time. But it has been under-appreciated until recently' [61].

Bamboo thrives in subtropical, tropical, and mild temperate regions across Africa, Asia, South America, and Oceania, mainly in Global South countries [62, 63]. Inhabitants have utilized bamboo for centuries, understanding its potential and limitations. Many local artisans are skilled in bamboo construction, requiring minimal training for new projects. Bamboo is an excellent lightweight structural material with high tensile strength, comparable to mild steel, and flexural strength ranging from 60 to 160 MPa, higher than many commercial woods [64, 65]. Its optimal thermal, acoustic, and tensile qualities, combined with effective seismic resistance due to its low mass, make it a favourable construction material in the Global South [66].

Figure 8.10 The arc at green school. Reproduced with permission from Tommaso Riva (2021), *ArchDaily*.

Additionally, bamboo's rapid growth and maturation, typically within 3–5 years, make it a sustainable resource, producing more green building materials and sequestering more carbon than typical tree forests [67]. Studies show bamboo forests can sequester seventeen times as much carbon as tree forests [68]. With around 1,500 bamboo species, only 20 to 30 are commonly used in construction, such as Moso bamboo (East Asia), Guadua bamboo (America), giant bamboo (Southeast Asia), and Oreobambos (Africa) [4]. In traditional buildings, bamboo can be used whole, split, pressed flat, or woven into mats, with joints made by tying ropes or incising the bamboo, showcasing its versatility and practicality [69].

Hardy's team merged local vernacular knowledge and craftsmanship with modern engineering to create the bamboo building, 'The Arc at Green School', an educational facility in Bali, Indonesia, completed in 2021. This unique structure occupy 760m2 space and features intersecting 14-m-tall bamboo arches that extend 19 m, interconnected by anticlastic gridshells, which gain strength from curving in opposite directions (Figure 8.10). The design enhances natural ventilation, essential for tropical climates, and allows for ample daylight, thanks to the large floor areas and elevated roof space.

One of the primary design challenges was to create a large, unobstructed space using minimal materials. The design drew inspiration from natural systems, particularly the way human ribcages are supported by the tension from an outer layer of muscle and skin. Project engineer Neil Thomas from Atelier One explained, 'The gridshells use shape stiffness to form the roof enclosure and provide buckling resistance to the parabolic arches. The two systems [gridshell and arches] together create a unique and highly efficient structure, able to flex under load, allowing the structure to redistribute weight, easing localised forces on the arches' [70]. The story of Green Village in Bali exemplifies how traditional materials and techniques from the Global South can create sustainable, innovative buildings. By embracing bamboo and combining it with contemporary design techniques, the Hardys and their team have crafted an environmentally friendly community deeply rooted in local culture and practices. This approach demonstrates the potential for sustainable building practices that honour and utilize local resources, offering a compelling model for communities worldwide. The creative blend of vernacular and indigenous knowledge with a modern engineering approach enables local craftsmen, architects, and engineers to construct buildings with low embodied carbon using materials suited to local climate conditions.

In 2004, the International Organization for Standardization (ISO) created standard 22157 to determine the mechanical properties of bamboo, including bending, compression, tension, and shear [9]. However, till now, Europe and other Global North countries still lack standardized regulations for bamboo as a building material, hindering its widespread adoption [65]. Meanwhile, in the Global South, the first building code for bamboo's application was introduced in India in 1994, '13985: Specification for bamboo mat board for general purposes', followed by the Chinese regulation 'GB/T 15780: Testing methods for physical and mechanical properties of bamboo' in 1995 and 'GB/T 2690 Bamboo timber' in 2000 [9]. By learning from the Global South's extensive experience with bamboo, the Global North can learn and adopt this sustainable material for modern construction, bridging cultural and environmental divides.

8.4 Conclusion

Recognizing the sustainability, economy, and intelligence of knowledge and technology from the Global South is crucial. Conventional approaches in the Global North could benefit from integrating these innovative practices. Embracing low-tech and low-process methods to reduce embodied carbon in buildings not only supports sustainability but also celebrates cultural heritage and practical wisdom. Key principles from the Global South include designing systems rather than individual parts, building for longevity, and using localized, low-cost materials and technologies. These principles, once abundant in traditional Global North architecture, have been devalued by modern technical advancements. In the Global North, low-impact living often evokes images of off-grid lifestyles with solar panels for electricity. While valid, this perception overlooks the potential of blending traditional and vernacular knowledge with low-cost and low-tech solutions. The question comes down to this: are solar panels the only solution for everyone and everywhere?

Respecting and building upon traditional and local knowledge allows each generation to enhance the wisdom of the past, rooted in local climatic conditions and culture. The success stories of Gando Primary School and Green Village illustrate the power of integrating local knowledge rather than applying uniform Global North solutions and demonstrate that sustainable building practices can be both technology smart and cost effective, resulting in buildings that are culturally rich, beautiful, and harmonious with their environment.

References

1. Wiik MK, Fufa SM, Kristjansdottir T, Andresen I. Lessons learnt from embodied GHG emission calculations in zero emission buildings (ZEBs) from the Norwegian ZEB research centre. Energy and Buildings. 2018 Apr; 165: 25–34.
2. Rapoport A. House form and culture. Hoboken, NJ: Prentice Hall; 2002.
3. Oldershaw B. Blackfeet American Indian women: builders of the tribe. Places. 1987; 4(1): 38–47.
4. Hungry Wolf A. The Blood People: a division of the Blackfoot Confederacy: an illustrated interpretation of the old ways [Internet]. New York: Harper&Row; 1977. Available from https://ehrafworldcultures.yale.edu/document?id=nf06-021
5. The United Nations. Sustainability [Internet]. [Accessed 15 Jul 2024]. Available from https://www.un.org/en/academic-impact/sustainability
6. U.S Environmental Protection Agency. Green building tools for tribes [Internet]. 2024 [Accessed 11 Jun 2024]. Available from https://www.epa.gov/green-building-tools-tribes
7. Hu M. Embodied carbon emissions of the residential building stock in the United States and the effectiveness of mitigation strategies. Climate. 2022 Sep 20; 10(10): 135.
8. UNEP. Global status report for buildings and construction: towards a zero-emission, efficient and resilient buildings and construction sector [Internet]. Nairobi; 2021 [Accessed 13 Sep 2022]. Available from https://globalabc.org/sites/default/files/2021-10/2021%20Buildings-GSR%20-%20Executive%20Summary%20ENG.pdf
9. Amede EA, Hailemariama EK, Hailemariam LM, Nuramo DA. A review of codes and standards for bamboo structural design. Advances in Materials Science and Engineering. 2021 Oct 31; 2021: 1–9.
10. Hrasnica M, Medic S. Seismic response of unreinforced masonry buildings from 1950's. In: 1st Croatian Conference on Earthquake Engineering [Internet]. University of Zagreb Faculty of Civil Engineering; 2021 [Accessed 9 Jun 2024], pp. 1137–48. Available from https://crocee.grad.hr/event/1/contributions/110/

11. Yu CR, Liu X, Wang QC, Yang D. Solving the comfort-retrofit conundrum through post-occupancy evaluation and multi-objective optimisation. Building Services Engineering Research and Technology. 2023 Jul; 44(4): 381–403.
12. Worstall T. Why so much US infrastructure and housing is lousy, built with shoddy materials [Internet]. 2014 [Accessed 9 Jun 2024]. Available from https://www.forbes.com/sites/timworstall/2014/06/27/why-so-much-us-infrastructure-and-housing-is-lousy-built-with-shoddy-materials/?sh=217506df7428
13. Churkina G, Organschi A, Reyer CPO, Ruff A, Vinke K, Liu Z, et al. Buildings as a global carbon sink. Nat Sustain. 2020 Apr; 3(4): 269–76.
14. Hamiltton I, Rapf O, Kockat D, Zuhaib D. 2021 Global status report for buildings and construction [Internet]. United Nations Environmental Programme; 2021 Oct. Available from https://www.unep.org/resources/report/2021-global-status-report-buildings-and-construction
15. Göswein V, Arehart J, Pittau F, Pomponi F, Lamb S, Zea Escamilla E, et al. Wood in buildings: the right answer to the wrong question. IOP Conf Ser: Earth Environ Sci. 2022 Sep 1; 1078(1): 012067.
16. Oliver P. Built to meet needs: cultural issues in vernacular architecture [Internet]. London: Routledge; 2007 [Accessed 23 Nov 2022]. Available from https://www.taylorfrancis.com/books/9781136424052
17. Curran MA. Biobased materials. In: Kirk-Othmer Encyclopedia of Chemical Technology [Internet]. Chichester, UK: John Wiley & Sons; 2010. Available from https://cfpub.epa.gov/si/si_public_record_report.cfm?Lab=NRMRL&dirEntryId=231873
18. Pomponi F, Hart J, Arehart JH, D'Amico B. Buildings as a global carbon sink? A reality check on feasibility limits. One Earth. 2020 Aug; 3(2): 157–61.
19. Lal R, Negassa W, Lorenz K. Carbon sequestration in soil. Current Opinion in Environmental Sustainability. 2015 Aug; 15: 79–86.
20. Hu M. Exploring low-carbon design and construction techniques: lessons from vernacular architecture. Climate. 2023; 11(8): 165.
21. Qu JJ, Cheng GD, Zhang KC, Wang JC, Zu RP, Fang HY. An experimental study of the mechanisms of freeze/thaw and wind erosion of ancient adobe buildings in northwest China. Bull Eng Geol Environ. 2007 May; 66(2): 153–9.
22. Silveira D, Varum H, Costa A, Martins T, Pereira H, Almeida J. Mechanical properties of adobe bricks in ancient constructions. Construction and Building Materials. 2012 Mar; 28(1): 36–44.
23. Webster FA, Tolles EL. Earthquake damage to historic and older adobe buildings during the 1994 Northridge, California earthquake. In 12WCEE 2000. 1999.
24. Morris J, Blier SP. Butabu: adobe architecture of West Africa. Princeton, NJ: Princeton Architectural Press; 2004.
25. Ramakrishnan S, Loganayagan S, Kowshika G, Ramprakash C, Aruneshwaran M. Adobe blocks reinforced with natural fibres: a review. Materials Today: Proceedings. 2021; 45: 6493–9.
26. Bouguerra A, Ledhem A, de Barquin F, Dheilly RM, Quéneudec M. Effect of microstructure on the mechanical and thermal properties of lightweight concrete prepared from clay, cement, and wood aggregates. Cement and Concrete Research. 1998 Aug; 28(8): 1179–90.

27. East East. The mid brick manual [Internet]. [Accessed 17 Jul 2024]. Available from https://easteast.world/posts/453
28. Bahobail MA. The mud additives and their effect on thermal conductivity of adobe bricks. Journal of Engineering Sciences. 2012 Jan 1; 40(1): 21–34.
29. Onyegiri I, Ugochukwu IB. Traditional building materials as a sustainable resource and material for low cost housing in Nigeria: Advantages, challenges and the way forward. International Journal of Research in Chemical, Metallurgical and Civil Engineering. 2016; 3(2): 247–52.
30. Ejiga O, Paul O, Cordelia O. Sustainability in traditional African architecture: a springboard for sustainable urban cities. In Sustainable futures: architecture and urbanism in the Global South, Kampala, Uganda, 27–30 June 2012, pp. 97–105.
31. Kubba S. Green building materials and products. In: Handbook of green building design and construction, 2nd ed; Oxford, UK: Butterworth-Heinemann, 2017, pp. 257–351.
32. Tlaiji G, Biwole P, Ouldboukhitine S, Pennec F. A mini-review on straw bale construction. Energies. 2022 Oct 23; 15(21): 7859.
33. Goodhew S, Griffiths R. Sustainable earth walls to meet the building regulations. Energy and Buildings. 2005 May; 37(5): 451–9.
34. Marques B, Tadeu A, Almeida J, António J, de Brito J. Characterisation of sustainable building walls made from rice straw bales. Journal of Building Engineering. 2020 Mar; 28: 101041.
35. Weyant C, Athalye V, Ragavan S, Rajarathnam U, Lalchandani D, Maithel S, et al. Emissions from South Asian brick production. Environ Sci Technol. 2014 Jun 3; 48(11): 6477–83.
36. Singh S, Maiti S, Bisht RS, Panigrahi SK, Yadav S. Large CO_2 reduction and enhanced thermal performance of agro-forestry, construction and demolition waste based fly ash bricks for sustainable construction. Sci Rep. 2024 Apr 10; 14(1): 8368.
37. Nath AJ, Lal R, Das AK. Fired bricks: CO_2 emission and food insecurity. Global Challenges. 2018 Apr; 2(4): 1700115.
38. Sarkar C, Sinha V, Sinha B, Panday AK, Rupakheti M, Lawrence MG. Source apportionment of NMVOCs in the Kathmandu Valley during the SusKat-ABC international field campaign using positive matrix factorization. Atmos Chem Phys. 2017 Jul 4; 17(13): 8129–56.
39. Ali AAM. Environmental impacts assessment of rice straw brick as a substitutional sustainable building material in Assiut University Hospital Clinic. Journal of Advanced Engineering Trends. 2021 Jul 1; 41(2): 247–59.
40. Khurshid I, Ahmad S, Nawaz R, Arshad M, Dar MEUI, Imran M, et al. Development of fire bricks from organic waste: an eco-friendly energy solution. Appl Ecol Env Res. 2018; 16(4): 3919–32.
41. Salim NSA, Kadir AA, Noor NM. Investigation of indoor air quality by incorporating sewage sludge waste into fired clay brick. IOP Conf Ser: Earth Environ Sci. 2023 Jul 1; 1216(1): 012007.
42. Mohajerani A, Ukwatta A, Jeffrey-Bailey T, Swaney M, Ahmed M, Rodwell G, et al. A proposal for recycling the world's unused stockpiles of treated wastewater sludge (biosolids) in fired-clay bricks. Buildings. 2019 Jan 5; 9(1): 14.

43. Srisuwan A, Sompech S, Saengthong C, Thaomola S, Chindraprasirt P, Phonphuak N. Preparation and properties of fired clay bricks with added wood ash. J Met Mater Miner. 2020 Dec 22; 30(4): 84–9.
44. Kadir AA, Sarani NA, Leman AM. Testing on building material using waste material in fired clay brick. MSF. 2014 Aug; 803: 330–6.
45. Walker R, Pavía S. Moisture transfer and thermal properties of hemp–lime concretes. Construction and Building Materials. 2014 Aug; 64: 270–6.
46. Fernando PR, Krishanth S, Rathnayake NB, Welarahne SA. Manufacturing, Physical and chemical characterization of fire clay brick value added with cow dung ash. AJMSP. 2019; 4(1): 32.
47. Kulshreshtha Y, Vardon PJ, Meesters G, Van Loosdrecht MCM, Mota NJA, Jonkers HM. What makes cow-dung stabilised earthen block water-resistant. In Barcelona, Spain; 2022 [Accessed 17 Jul 2024], pp. 540–8. Available from https://www.scientific.net/CTA.1.540
48. Hořínková D. Advantages and disadvantages of modular construction, including environmental impacts. IOP Conf Ser: Mater Sci Eng. 2021 Nov 1; 1203(3): 032002.
49. Gwiazda M. Modular designs at the early Byzantine pilgrimage site of Philoxenite, Egypt. J Roman Archaeol. 2023 Jun; 36(1): 196–214.
50. Oleson JP. The modular planning of Roman fortifications in the Near East: principles and process. In: The Socio-economic History and Material Culture of the Roman and Byzantine Near East [Internet]. Gorgias Press; 2017 [Accessed 8 Jul 2024], pp. 237–72. Available from https://www.degruyter.com/document/doi/10.31826/9781463237738-012/html
51. Knapp RG. China's old dwellings. Honolulu: University of Hawaii Press; 2000.
52. Kyakula M, Gombya I. Suitability of bamboo for construction and environmental preservation. Journal of Civil Engineering Research and Practice. 2008 Jul 16; 5(1): 43–51.
53. Rashid M, Ara DR. Modernity in tradition: reflections on building design and technology in the Asian vernacular. Frontiers of Architectural Research. 2015 Mar; 4(1): 46–55.
54. Munaro MR, Tavares SF, Bragança L. Towards circular and more sustainable buildings: A systematic literature review on the circular economy in the built environment. Journal of Cleaner Production. 2020; 260: 121134.
55. Minunno R, O'Grady T, Morrison GM, Gruner RL. Investigating the embodied energy and carbon of buildings: a systematic literature review and meta-analysis of life cycle assessments. Renewable and Sustainable Energy Reviews. 2021 Jun; 143: 110935.
56. Knaack AM, Kurama YC. Behavior of reinforced concrete beams with recycled concrete coarse aggregates. J Struct Eng. 2015 Mar; 141(3): B4014009.
57. Weiler K, Gutschow N, editors. Authenticity in architectural heritage conservation: discourses, opinions, experiences in Europe, South and East Asia. Cham: Springer International Publishing Switzerland; 2017. (Transcultural research—Heidelberg studies on Asia and Europe in a global context).
58. Kéré F. Primary school in Gando, Burkina Faso. [Internet]. uma architecture & design. [Accessed 2024 Jun 11]. Available from https://www.uma-archdesign.com/primary-school-gando-burkina-faso-francis-kr

59. Block I. Diébédo Francis Kéré says school that launched his career is 'not a traditional African building' [Internet]. 2017 [Accessed 11 Jun 2024]. Available from https://www.dezeen.com/2017/10/17/movie-diebedo-francis-kere-gando-school-burkina-faso-interview-video/
60. 5 things to know about new running bamboo ordinance, effective Jan. 1 [Internet]. Fairfax County Government NewsCenter. 2022 [Accessed 13 Jun 2024]. Available from https://www.fairfaxcounty.gov/news/5-things-know-about-new-running-bamboo-ordinance-effective-jan-1#:~:text=Quick%20Summary,and%20suppress%20native%20plant%20species.
61. Anderson C. Detail: Bamboo structure of the arc at Green School Bali [Internet]. archello. 2023 [Accessed 13 Jun 2024]. Available from https://archello.com/news/detail-bamboo-structure-of-the-arc-at-green-school-bali
62. Agyekum K, Kissi E, Danku JC. Vengala. Scientific African. 2020 Jul; 8: e00424.
63. Vengala J, Mohanthy B, Raghunath S. Seismic performance of bamboo housing—an overview. In Proc. of World Bamboo Congress Vol 1 (Damyang, Korea); 2015. pp. 389–407.
64. Zhang K, Yu L, Dai F, Chen Y, Jiang Z, Wang Y, et al. Bamboo structure and its impact on mechanical properties: a case study of *Bambusa arundinaceae*. Forests. 2024 Apr 26; 15(5): 762.
65. Bamboo Imort Europe. The mechanical properties of bamboo [Internet]. [Accessed 13 Jun 2024]. Available from https://bambooimport.com/en/pages/the-mechanical-properties-of-bamboo
66. Mili M, Singhwane A, Hada V, Naik A, Nair P, Kumar Srivastava A, et al. Advances in bamboo composites for structural applications: a review. In: Bamboo—recent development and application [Working Title] [Internet]. IntechOpen; 2023 [Accessed 13 Jun 2024]. Available from https://www.intechopen.com/online-first/86596
67. Sharma B, Gatóo A, Ramage MH. Effect of processing methods on the mechanical properties of engineered bamboo. Construction and Building Materials. 2015 May; 83: 95–101.
68. Disén K, Clouston PL. Building with bamboo: a review of culm connection technology. Journal of Green Building. 2013 Sep; 8(4): 83–93.
69. Sassu M, De Falco A, Giresini L, Puppio M. Structural solutions for low-cost bamboo frames: experimental tests and constructive assessments. Materials. 2016 May 7; 9(5): 346.
70. Griffiths A. Ibuku completes 'unprecedented' bamboo building in the Balinese jungle [Internet]. dezeen. 2021 [Accessed 13 Jun 2024]. Available from https://www.dezeen.com/2021/07/24/ibuku-the-arc-bamboo-jungle-bali-architecture/

Chapter 9
Equity, embodied carbon proposals, and road to justice

9.1 Looking back, looking ahead

The book has been an effort to chart a new course in understanding embodied carbon issues and related Global North–South inequity. The previous chapters argued that the greatest problems threatening the viability of a future North–South embodied carbon policy cooperation are persistent global inequality, an ongoing development crisis in much of the South, and starkly imbalanced suffering from, responsibility for, and action on carbon emission and resource conservation. Without addressing these, the world will continue to experience increasing carbon emissions and move ever further away from carbon neutrality. The root of the problem, the author argues, is the ignorance of the accurate counting of embodied carbon and bias towards continuous growth and technological advancement, which are aided by the unilateral carbon neutral requirements developed and dominated by Global North economic institutions like the European Union and Organization for Economic Co-operation and Development.

Part I, Looking back and perception, comprises Chapters 1, 2, and 3. Chapter 1 takes readers on a journey through the historical evolution of embodied carbon, revealing the origins and development of this crucial concept in the building and construction sector and exploring how measurement methods have evolved over time. Chapter 2 dives deep into the various definitions, components, and accounting methods of embodied carbon, unravelling the complexities of the environmental product declaration market. Chapter 3 makes a compelling case for the critical connection between embodied carbon and climate change, highlighting its significance in climate action and showcasing real-world impacts.

Embodied carbon, the total greenhouse gas emissions associated with the production, transportation, and assembly of building materials, is a critical aspect of a building's overall carbon footprint that cannot be ignored. Embodied carbon can account for a substantial proportion of a building's

Embodied Carbon. Ming Hu, Oxford University Press. © Ming Hu (2025). DOI: 10.1093/9780198944485.003.0009

total emissions, sometimes up to 80% or more over the building's life cycle [1]. Focusing solely on operational carbon—emissions from the building's use, such as heating, cooling, and lighting—neglects this significant source of emissions, leading to an incomplete understanding of the building's environmental impact. Addressing embodied carbon ensures a more holistic approach to emission reductions. This comprehensive view is essential for assessing the true environmental impact of buildings and infrastructure projects.

After clarifying the importance and necessity of addressing embodied carbon at a global scale in the built environment, Part II offers a comprehensive view of the complexities and hidden biases surrounding embodied carbon, providing valuable insights for more equitable and sustainable practices. This section includes Chapters 4 through 6. Chapter 4 begins by emphasizing that the distribution of embodied carbon is not limited to the industry alone but is deeply intertwined with the socioeconomic fabric of our societies. It argues that improving efficiency and being environmentally conscious alone cannot solve the problem, and the industry cannot be the sole party responsible for reducing embodied carbon. This chapter explores the disparate realities of urban development, highlighting how marginalized communities often bear the brunt of environmental burdens and impacts. This sets the stage for a deeper examination of the systemic biases that influence our approaches to sustainability.

Building on this foundation, Chapter 5 addresses the Global North bias, analysing how it distorts our understanding of the world and examines its far-reaching effects on knowledge production, policymaking, and resource allocation in sustainability. By scrutinizing the assumptions and frameworks that prioritize the interests and perspectives of developed nations, this chapter reveals the inequities embedded in our current sustainability efforts. This exploration of bias transitions seamlessly into the next discussion on technological influences.

Chapter 6 then discusses the high-tech bias and its implications for the built environment and the circular economy, shedding light on technology-driven inequalities and the role of advanced technologies in perpetuating environmental disparities. It underscores how reliance on high-tech solutions can often overlook simpler, more accessible approaches that could benefit a broader segment of society. This chapter culminates the discussion by connecting technological and socioeconomic biases, highlighting the need for a more inclusive and equitable approach to addressing embodied carbon.

The last part, Part III, focuses on understanding current global efforts and potential paths forward to address embodied carbon inequality. Chapter 7 provides a comprehensive view of current embodied carbon emission-related regulations and policies and explains the pros and cons. Chapter 8 introduces the proposed low-tech and low-process approach that extracts knowledge from traditional and indigenous design and construction. Finally, this led us to an investigation in this chapter, Chapter 9, to understand what experts say about the appropriate approach, mindset, and solutions. This chapter is drawn from multidisciplinary perspectives since embodied carbon emission is not only an environmental issue, it is more of an economic and human development issue. In Section 9.2, we can observe the opinions of the experts from different fields to create a new roadmap framework (Section 9.4) of a comprehensive approach to reset the current imbalance and create more equitable solutions to reduce embodied carbon emission and related impact on planetary health.

9.2 What did experts say?

9.2.1 Sociologists

The field of sociology has been pivotal in shedding light on the climate change disparities between the Global North and South. A seminal work in this area is *A Climate of Injustice: Global Inequality, North-South Politics, and Climate Policy* by Dr. J. Timmons Roberts and Bradley C. Parks, published in 2007 [2]. This influential book dives deep into the heart of climate injustice, highlighting how poorer countries in the Global South suffer the most from climate change despite contributing the least to global greenhouse gas emissions.

Roberts and Parks vividly illustrate the historical and ongoing exploitation that has intensified these inequalities. They discuss the stark contrast between production-based and consumption-based emissions (Chapter 4). Developing countries, particularly in the Global South, often have high production-based emissions due to their role in manufacturing and resource extraction for global markets. Yet, these emissions are largely driven by the consumption habits of the Global North. The book brings to life specific examples, like the industrial sectors in China and India, which produce goods for export but bear the brunt of environmental and health costs from high emissions. They argue that international climate policies focusing solely on production-based emissions overlook historical contexts, unfairly penalizing developing nations while letting developed countries sidestep responsibility for their consumption-driven emissions. Instead, they advocate for equitable accounting methods that consider both production- and consumption-based emissions.

Roberts and Parks propose powerful strategies to combat climate injustice rooted in global inequality, emphasizing two pillars: global cooperation and fairer emission policies. They assert that Global North countries must acknowledge their historical responsibility for cumulative carbon emissions, take greater accountability for their current consumption patterns, and support developing countries in their transition to low-carbon economies with financial and technological aid. On the policy front, they recommend adopting a framework that recognizes the differing capacities and contributions of various countries to the climate crisis. This includes implementing fair carbon pricing mechanisms that account for both production- and consumption-based emissions, ensuring that Global North developed countries are accountable for their consumption-driven emissions, even when production occurs elsewhere. They also suggest border carbon adjustments to prevent carbon leakage and foster equitable global climate action.

Roberts and Parks paint a vision of a fairer, more effective response to climate change, significantly influencing embodied carbon policy by highlighting the need for a holistic approach that incorporates both historical emissions and consumption patterns. By integrating these measures and enhancing international cooperation and governance structures, they propose a path forward that prioritizes equity and justice, addressing the needs and rights of the world's most vulnerable populations. Their core argument that global climate change is fundamentally an issue of injustice and inequality, with poor countries suffering disproportionately despite contributing little to the problem, has been highly influential in framing climate change as an ethical issue of global justice. Their book is widely cited and has introduced these concepts to other disciplines, serving as a theoretical foundation for fields such as science, engineering, and technology. Their work is a compelling call to action for a more just and sustainable future.

9.2.2 Economists

Any social issue is always closely intertwined with economic challenges. Dr. Dieter Helm, a prominent British economist specializing in climate change policy, advocates for comprehensive and pragmatic approaches to decarbonization. In many of his publications, he has been highly critical of the European Union's progress over the past thirty years in mitigating climate change, particularly since the United Nations Conference on Environment and Development held in Rio de Janeiro in 1992. In his book *Net Zero: How We Stop Causing Climate Change* published in 2020, at the introduction, he stated:.

> Thirty years ago, world leaders vowed to address the new great -global warming. Thirty years later, the concentration of CO2 in the atmosphere continues to grow unabated . . . . The last 30 years have been the golden age of the fossil fuels. Far more fossil fuels have been burnt in the past 30 years than in the entire nineteenth century. [3]

He argues that the fundamental flaw over the past thirty years has been viewing the problem from the wrong perspective—akin to looking down the wrong end of a telescope. Instead of placing the blame solely on world leaders, corporations, or figures like Trump or the 'Chinese Communist Party's addiction to coal', he believes we should turn our focus inward and examine the consumption behaviours in EU-developed countries. Our long-held misguided approach has led to a failure to address climate change effectively, as the past thirty years have focused on achieving net-zero emissions and reducing production-based emissions on a territorial basis. The solution has been seen as merely switching production from high to low carbon within specific regions that adopt carbon reduction targets. Policymakers, practitioners, and researchers have mistakenly assumed that changing producer behaviour would solve the problem. However, the central flaw lies in this starting point. The real starting point should be 'consumption and us, the consumers'. Dr. Helm reminds us that all the products made with fossil fuels—and thus causing carbon emissions—are produced for us. Therefore, the true polluters are not just the producers, but us, the consumers, particularly those in Global North developed countries with high purchasing power.

He further explained that the fundamental missing piece in the current global economy is the cost of net zero or decarbonization, where consumers do not face higher costs caused by their own lifestyle or consumption behaviour. In his most recent book, *Legacy: How to Build the Sustainable Economy*, Dr. Helm points out:.

> The harsh reality is that we, the polluters, are living beyond our sustainable means. We are not paying for the carbon (and multiple other forms of) pollution we are causing. If we were, prices would be higher and living standards would be affected. We do not want to vote to pay for the damage we are causing. We are not living in a sustainable economy. [4]

Dr. Helm's proposal on carbon emission reduction and decarbonization comprises several components: global coordination, environmental and economic policy integration, technological innovation, and moral principles. For global coordination, while he acknowledges the necessity of individual countries taking unilateral steps towards decarbonization, he believes these efforts must be part of a broader, coordinated global strategy. Unilateral actions can serve as important catalysts and demonstrate leadership, but they must be complemented by multilateral agreements and policies to be truly effective. Another way to say this is that the previous and current universal carbon emission cuts are not practical and achievable.

For those passionate about sustainability, Helm's perspective on *integrating environmental and economic policies* offers a compelling narrative. He argues that our current economic model is fundamentally unsustainable and requires a complete overhaul to tackle environmental degradation and climate change effectively. According to Helm, we need well-designed energy policies and regulations that pave the way for a low-carbon economy. He criticizes short-term, fragmented approaches that fall short of addressing the systemic changes necessary for true decarbonization. Helm highlights the critical role of government in promoting sustainability through robust regulations and policy frameworks. He believes that stronger environmental regulations and incentives are essential for encouraging businesses to adopt sustainable practices. This approach stands in contrast to classical economic views that oppose trade restrictions, instead aligning with the intellectual heritage of Herman Daly's ecological economics. In Chapter 7, we discuss how combining environmental and tax policies could offer tailored solutions adaptable to both Global North and South countries. This integrated approach provides the flexibility needed to address diverse environmental and economic challenges across different regions, reinforcing Helm's vision for a sustainable future.

For *technological innovation*, Helm emphasizes the role of technological innovation in achieving decarbonization goals. He argues that investment in research and development of clean technologies is crucial, requiring national initiatives and international collaboration. He strongly believes in technical solutions yet did not specify any particular ones for building and construction sector.

Lastly, but most importantly, Dr. Helm's work extensively addresses *the moral and ethical dimensions* of environmental policy and sustainability. He believes that addressing climate change is not only a scientific and economic challenge but also a moral imperative. He calls for equitable solutions that consider the needs and rights of all people, especially those who are least responsible for causing climate change but most affected by its impacts, such as citizens in Global South developing countries. In his book, he suggests that citizens in Global North countries (e.g. the UK) get a complete picture of consumption-based emissions by writing down a carbon diary from the toilet paper to travel for work. The point is not to make people feel guilty; rather it is to reveal what the world will have to look like if the world wants to decarbonize. Helm claims, 'We are becoming addicted to a way of living in which flying is regarded as essential by many and an aspiration for most.' He not only highlights how unsustainable the lifestyles in the Global North have become; he also points out that some of the actions of climate change activists are not sustainable. In his book *Net Zero*, he mentions:.

> Even Greta Thungerg, in her noble efforts to get world leaders to take climate change seriously, had a support team flying backward and forwards to the 2019 UN summit on climate change as she sailed in a 'zero carbon' yacht. They had, it was claimed, no option. [3]

He highlights the critical importance of intergenerational equity, ethical stewardship, precautionary measures, and fairness in tackling climate change and managing natural resources. He champions environmental policies that equitably distribute the costs and benefits of sustainability initiatives, ensuring that no group or community is unfairly burdened. Helm insists that these policies should be transparent, accountable, and inclusive, with public participation playing a key role. This approach ensures that environmental decisions are made fairly and align with society's moral values, reflecting a commitment to just and ethical stewardship of our planet.

9.2.3 Engineers and technologists

Dr Francesco Pomponi is a prominent engineering researcher specializing in embodied carbon, focusing on sustainability science and its application in built environments. In an interview with the author, Dr Pomponi shared his views on the disparity between the Global North and South in terms of embodied carbon, drawing from his work in the Global South, particularly sub-Saharan Africa. He identifies two key issues in current carbon accounting practices: differing priorities between the Global North and South and how these priorities manifest. In the Global North, discussions on climate change and global warming are commonplace, influenced by the prevalent wealth in these societies. People often believe that electrifying their homes and adopting circular practices (recycling and reusing) can solve their contribution to climate change. However, these are oversimplifications. Scientists like Dr Pomponi understand the complexity of these issues. For example, simply electrifying homes does not automatically reduce energy consumption and carbon emissions without changes in consumer behaviour. Despite this, there is an almost obsessive focus on carbon in the Global North, with a prevailing narrative that reducing carbon emissions and achieving net-zero carbon will mitigate climate change.

In contrast, the situation in the Global South is different. In countries like South Africa, a reliable energy grid is not guaranteed. People struggle to maintain basic needs like refrigeration, bathing, and cooking, which are often taken for granted in wealthier countries. Here, the stability of the energy grid, regardless of its power source, is the priority, not fossil fuels. This disparity highlights a significant misunderstanding in the Global North, where there is often a failure to grasp the local challenges in the Global South that extend beyond carbon accounting. These challenges are deeply rooted in social, governmental, and sustainability issues, not just environmental concerns.

Dr Pomponi also pointed out the second issue: how carbon accounting is conducted in the Global South. There is a risk in applying a Global North perspective, especially within the built environment. This perspective tends to overlook vernacular construction methods, local materials, and traditional architecture. Instead, it focuses on carbon accounting solutions that heavily rely on multistory framed buildings made of concrete, steel, cement, and glass, with minimal use of local materials, just like those built in the Global North. There is a notable lack of interest in developing datasets that reflect local building materials, stemming from an assumption that the best way to construct buildings is to follow modern methods used in the Global North. When some opinionated Global North scholars thought that people in Africa lived in mud huts without setting foot on the continent, they questioned the types of materials needed for assessment in Africa. The reality is that Africa has many local and low-carbon building materials, but the influence from the colonization period planted the wrong perception about their utility and value. High-tech solutions like artificial intelligence (AI) will enhance this misperception because the algorithms are developed in the Global North. As a result, local people receive the wrong information from these 'high-tech' solutions and aspire to build environments similar to those in the Global North, using carbon-intensive materials like glass, steel, and concrete.

This assumption and high-tech bias have led to significant embodied and operational carbon problems in our current built environments, as evidenced by the continued rise of carbon emissions since the Paris Agreement. Dr. Pomponi stated, 'The fact that we completely neglected vernacular architecture means that we're going towards a world dominated by mechanical heating and cooling of buildings.' This high-tech shift ignores the surrounding climate, traditional building practices, variations in building heights, and the importance of orientation in relation to natural elements like wind. Consequently, the most sustainable and cost-effective solutions rooted in local, traditional, and vernacular architecture are ignored and forgotten. For example, traditional wind catchers in Persia and small windows in certain parts of Italy designed to minimize heat intake during summer demonstrate the ingenuity of vernacular solutions. The high-tech bias, mentioned in Chapter 6, has worsened this ignorance and widened the equality gap. Another example of high-tech bias is the use of AI. AI algorithms have been increasingly scrutinized for biases, particularly in reflecting the perspectives of various demographic groups. The lack of diversity in the AI field, where most developers are white and male, can lead to biases being overlooked or not identified [5]. Dr. Pomponi highlighted this issue using his own experience in the Global South, stating, 'Using AI would just lead to the same unfortunate outcome a bit faster.'

To summarize, Dr Pomponi pointed out two main issues with current embodied carbon practices and approaches: first, carbon accounting practices are incomplete, and second, when carbon accounting is performed in the Global South, it is often done through the lens of the Global North. This approach overlooks local contexts and solutions that could be more sustainable and appropriate for the regions in question. Instead, evidence-based strategies derived from Global South knowledge (explained in Chapter 8) are essential for effectively reducing embodied carbon through policy.

9.2.4 Engineers and policymakers

Engineers' contributions and roles can extend beyond technical solutions; when engineer-trained policymakers hold administrative positions in government, they can shape key carbon and energy policies. One notable figure is Dr Clark Bullard, a professor at the University of Illinois. Dr. Bullard has dedicated his career to energy and water conservation policy. He served as the Director of the Conservation and Renewable Energy Policy Office at the

U.S. Department of Energy during the Carter Administration and participated in numerous Illinois task forces and commissions related to water and energy policy during the Thompson and Edgar Administrations. These experiences provided him with a deep understanding of policymaking. During an hour-long interview with the author, Dr. Bullard shared his experience and opinions on various topics, including addressing inequity issues and the importance of policy.

Dr. Bullard began by addressing embodied carbon regulation, which often involves economic strategies, such as taxation. Economists advocate for taxing undesirable activities, like energy consumption and carbon emissions. However, energy taxes disproportionately affect low-income families since they constitute a larger percentage of their income. Carbon taxes present similar challenges. Policies must ensure broad participation and fairness to address these equity issues while maintaining economic efficiency. Over forty years ago, Dr. Bullard proposed a global carbon trading policy during his tenure in the Carter Administration. This policy would allocate a specific carbon emission allowance per person annually, which could be traded. The concept aimed to balance equity and efficiency by allowing market mechanisms to handle trading while the initial distribution addressed equity. Although the idea faced resistance in the U.S., particularly due to concerns about international trade imbalances, it offered a framework for integrating equity into carbon reduction efforts. Specifically, under Dr Bullard's proposal, each person on Earth would be entitled to emit a certain number of tons of carbon annually. These allowances could be bought and sold, creating a market for carbon certificates. The general principle was that Americans, who have higher per capita emissions, should reduce their carbon footprint more than individuals in developing countries in the Global South. Each year, the number of certificates would decrease, with each certificate worth 90% of its value the following year. This system separated equity from efficiency: the market would determine the price of the certificate (efficiency), while the initial distribution of permits would address equity. For example, wealthier individuals, who typically have higher carbon emissions, would receive fewer certificates.

From Bullard's perspective, cap-and-trade policies, which separate discussions of efficiency and equity, offer a practical and fair approach. Even though the U.S. experienced success with cap-and-trade, such as the Acid Rain Program under the 1990 Clean Air Act Amendments, carbon cap-and-trade policies remain contentious. For example, California's system faced legal challenges over its classification as a tax but was ultimately upheld because purchasing emissions allowances is a voluntary business decision. The U.S.

lacks a national carbon cap-and-trade policy, leading to varied state-level programmes. Thirteen states, including California and New York, have implemented their own systems. In the absence of national policy, the federal government relies on infrastructure bills, tax credits, and grants to incentivize carbon reduction. While effective in the short term, this approach can become economically inefficient if overused.

Dr. Bullard believes a policy-driven approach should replace cash incentives in the U.S., as we cannot always pay people to do the right thing. Government spending plays a pivotal role in reducing carbon emissions through policy interventions. By investing in various sectors and implementing targeted policies, governments can effectively lower emissions. Research by Busch and colleagues suggests that economic policies, such as paying for net emission reductions at a district scale and sharing revenues with the national government, can significantly reduce emissions [6]. Evidence from countries with higher government spending in the Global South further supports the effectiveness of carbon emission-reduction policies [7]. Example policies from the Global North are exemplified by Scandinavian countries, where government expenditure as a percentage of GDP is significantly higher than that in the U.S. (36%), demonstrating the potential for aggressive carbon emission regulations funded by high taxation. Countries like Denmark, Norway, and Sweden rely on substantial tax revenues (50–60%) to support extensive welfare states and strict environmental policies [8]. If the U.S. wants to achieve similar results, a social consensus on higher taxation levels is required, which might be politically difficult.

In conclusion, integrating evidence-based strategies into policymaking, leveraging both economic policies and regulatory frameworks, is crucial for achieving significant reductions in embodied carbon. The experiences of researchers-turned-policymakers like Dr. Bullard underscore the importance of informed, equitable, and efficient policies in addressing global carbon challenges.

9.2.5 Synergies from different views

The synergies among these expert perspectives reveal a compelling convergence of themes. First and foremost, there's a universal call for global cooperation and equitable policies, with a clear recognition of the Global North's historical and current responsibilities. Experts advocate for a seamless integration of environmental sustainability with economic strategies, emphasizing the need for carbon pricing mechanisms, financial incentives for sustainable practices, and strong regulatory frameworks.

Technological innovation is another shared focus, but experts' opinions are different from popular high-tech-focused and growth-driven approaches. Instead, they emphasize the crucial role of leveraging local materials and traditional methods in the Global South. This approach not only aligns with the sociological emphasis on fair and contextually appropriate solutions but also ensures that embodied carbon is included and valued. Additionally, the moral dimensions of carbon policy are brought to the forefront, underscoring the ethical responsibility of the Global North to support or compensate the sustainable development in the Global South.

There's also a unanimous call for evidence-based and pragmatic approaches in policy formulation. Experts stress the importance of data-driven decisions and the necessity for transparent, accountable, and inclusive decision-making processes. By weaving together these perspectives, it becomes evident that addressing embodied carbon demands a multifaceted approach. This approach must integrate *global cooperation*, *equitable policies*, *technological appropriateness*, and a *strong ethical foundation*. These synergies offer a comprehensive framework for developing effective and just climate policies that address the needs and rights of all populations, particularly the most vulnerable.

9.3 Breaking the impasses

This final chapter is about building a long-term foundation for meaningful Global North–South environmental cooperation, with a focus on embodied carbon emissions. However, before suggesting the need for changes in the 'rules of the game' that govern the contemporary global political economy, it may be worthwhile to assess the urgency of addressing the embodied carbon issue.

According to the United Nations Environment Program Global Status Report, 'By 2060, the world is projected to add 230 billion m^2 (2.5 trillion ft^2) of buildings, equivalent to adding an entire New York City to the planet every 34 days for the next 40 years' [9]. Construction activities and urbanization are mainly stagnant in the Global North's developed countries. Among G20 countries, only the U.S. has shown consistent growth in construction. In contrast, many Global South countries have experienced sustained population growth and increased construction activities [10]. These trends in the Global South are expected to continue, with global construction output projected to rise by over 70%, reaching US$15 trillion, primarily in these countries. According to UN-Habitat, the floor area within cities in low-income countries of the Global South is estimated to increase by 141% within fifty years from 2020 levels. This rapid urbanization will necessitate massive new construction to house the influx of urban populations and build supporting infrastructure, all of which will increase embodied carbon emissions globally. Regions with rising construction activities and limited access to low-carbon materials risk locking in high embodied carbon and operational emissions for an extended period.

The fundamental political question is who will be responsible for the bulk of future global embodied carbon emissions. As we saw in Chapter 4, the current accumulated stock of carbon emissions in the atmosphere is largely the responsibility of rich, industrialized Global North countries. However, future emissions will primarily increase in the Global South, driven by population and economic growth. Population growth is an unstoppable factor due to various reasons, including improved human health and longevity. Similarly, the economic growth of developing countries is inevitable, as all humans deserve to have their basic needs met. Together, these factors are set to increase the Global South's carbon emissions from 31% of the total in 1990 to 60% in 2030 [11].

We must, therefore, ask several important questions: How can Global North and South countries return to the cooperation table regarding embodied carbon emissions? What if the Global South demands significant changes in the Global North's agenda on embodied carbon, potentially leading to stagnation in living standards? Conversely, what if the Global North demands a firm commitment from the Global South to reduce embodied carbon emissions that can be detrimental to improving the living standards and economic growth of those poor countries? While there is much uncertainty surrounding these questions, one thing is clear: a global climate accord without long-term participation from both the Global North and South will not bring about significant change [2].

So, what are the long-term prospects for a meaningful, equitable global approach to embodied carbon? Many insights can be drawn from experts' opinions, each bringing knowledge from different fields. Extracted the information from the experts' opinions mentioned in Section 9.2, this chapter outlines three key principles, illustrated in Figure 9.1. *First*, polluters pay:

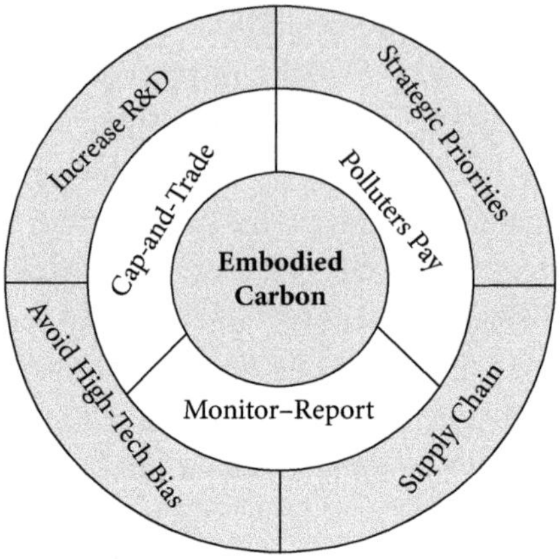

Figure 9.1 Roadmap action framework.

the polluters are the consumers who generate the demand, and the mechanism will combine a carbon tax and carbon tariff. This way, the demand for consumption can be regulated and reduced. *Second*, setting up a cap-and-trade with carbon allowance and credit exchange involves establishing a carbon allowance/quota and allowing carbon credit exchanges. Most Global North countries have already reached their carbon allowance, so they must first undergo degrowth and then offset any future emissions deemed necessary. *Third*, create a robust monitoring and reporting system: an efficient and trustworthy system is crucial for tracking progress and ensuring accountability. These principles provide a foundation for addressing the complex challenge of global embodied carbon emissions equitably and effectively. Those principles together create an integrated platform utilizing both environmental and economic policy to create a sustainable and equitable pathway.

9.4 Upgrade the Development Framework

Now that we have a platform supported by three key principles, we can develop a roadmap and action framework. As illustrated in Figure 9.1, the proposed framework offers a more equitable approach to controlling embodied carbon by better considering the differing needs and dynamics of the Global North and South. It takes into account the entire building supply chain and carbon emissions related to material manufacturing and building construction.

Additionally, it integrates climate adaptation and inclusion as central elements, providing an equitable path forward. The framework comprises three principles (identified in Section 9.3) that cut across the following four action areas:.

Action Area 1—Strategic priorities encompass national, regional, and local development planning processes that outline development objectives, prioritize them, and guide their implementation. This action area underscores the necessity of incorporating inclusive embodied carbon reduction strategies within the buildings and construction sector at the highest levels of governance. Doing so provides a clear direction and mandate to line ministries and local governments, ensuring effective implementation.

Action Area 2—Increase research and development (R&D) in environmental science: integrate vernacular and local knowledge. It highlights the importance of leveraging traditional and indigenous practices, which are often more sustainable and better suited to local environments. Investing in R&D that values these methods can uncover low-carbon solutions that modern science frequently overlooks, promoting sustainability and preserving cultural heritage.

Action Area 3—Avoid high-tech bias and promote low-tech solutions: prioritize simple, sustainable technologies that are accessible and effective, particularly in the Global South. By emphasizing low-tech approaches, we can address local needs more equitably, reduce reliance on complex and expensive technologies, and leverage traditional methods that are often more environmentally friendly and culturally appropriate.

Action Area 4—Construction supply chain: includes raw material extraction, system, appliance and material manufacturing, commercialization, importation, and the norms regulating the supply chain. This action area is critical to reducing the building's embodied carbon by acting upstream, implementing circularity principles and sustainable production as well as ensuring equity and well-being for the industry workforce and the community.

The four action areas were defined to cover the entire building and construction supply chain and life cycle to ensure comprehensive and equitable embodied carbon reduction action. In addition, the four action areas can be set so that each area can be considered independently and adopted by a group of stakeholders. In the following sections, we will focus on action areas 2 and 3 to demonstrate their importance and how to engage stakeholders.

9.4.1 Advancing environmental science R&D through local database development

The need for R&D in environmental science is critical, particularly for integrating vernacular and local knowledge to address the embodied carbon challenge. Currently, the lack of a comprehensive database makes embodied carbon results unreliable and potentially leads to misguided policies. This data deficiency is especially pronounced for low-embodied carbon materials used in the Global South, such as adobe brick. As Dr Pomponi pointed out, viewing this global issue predominantly through a Global North lens is a significant part of the problem. Addressing this requires overcoming specific data collection challenges, including comparability, completeness, and transparency.

Comparability in assessments means that different evaluators should use the same criteria to ensure accurate comparisons. A significant barrier to this is the variability in completeness. For example, one assessment might include certain life-cycle stages that another omits, leading to differing results—not because one building is inherently better, but due to subjective choices in data inclusion. To overcome this, developing additional guidance, standards, and data schemes is essential. These measures aim to ensure that assessments clearly state what is included and exclude nothing significant (over 5% of the building's cost or volume), thereby enhancing comparability and completeness.

Another challenge is the type of data used to support assessments. For instance, Environmental Product Declarations (EPDs) for materials like steel can vary significantly. One assessment might use data from a steel production process with a high carbon footprint (e.g. basic oxygen furnaces), while another uses data from a process with a lower footprint (e.g. electric arc furnaces). Large steel mills using basic oxygen furnaces (BOFs) primarily process raw materials, leading to lower recycled content and higher carbon footprints. In contrast, smaller factories using electric arc furnaces (EAFs) melt scrap iron and steel, achieving high recycled content and lower carbon footprints.

The differences in steel-making methods between the Global North and South are notable. For instance, in 2022, India, the world's second-largest steel producer, predominantly used BOFs (46%) while EAFs accounted for only 23%. [12] In China, BOFs account for 90% of output, whereas less than 30% of U.S. steel is made this way, according to a 2021 Rocky Mountain Institute report [13]. Investing in R&D to improve steel-making technology and processes in Global South countries could significantly reduce embodied carbon. This approach should also integrate traditional and indigenous practices that have proven sustainable over centuries.

9.4.2 Reviving vernacular wisdom in the green transition

For any new green initiative to successfully adapt to climate change, local wisdom—rooted in a deep understanding of the land, nature, and culture—must be prioritized over shiny new technological advancements. In 2019, the European Commission launched the European Green Deal, injecting billions of euros into the green transition. By September 2021, they had also created the New European Bauhaus (NEB), inspired by the early twentieth-century German arts and crafts movement. NEB is built on three principles: sustainability, inclusion, and beauty, aiming to transform Europe's green renovation into a cultural movement by bridging science, technology, art, and culture [14]. However, there is concern that NEB could follow in the old Bauhaus movement's footsteps. Despite its revolutionary intent, the Bauhaus movement has been reduced to a style characterized by 'houses with lots of glass and shining metal' [15]. This trend led to the proliferation of glass-box, high-rise buildings in Asia and the Middle East, which are energy and carbon-intensive. The NEB initiative promises a sustainable and inclusive future, encouraging dialogue across cultures, disciplines, genders, and ages. But the detailed action plans in the NEB concept paper presented to the President of the European Commission fell short of truly integrating vernacular knowledge. The paper's statement that 'both frontier technologies and vernacular techniques must be equally embraced' positions vernacular techniques as a separate category from advanced technologies. In reality, vernacular knowledge—developed through a deep understanding of the symbiotic relationship between built and natural environments—should play a leading role in climate adaptation and mitigation [16].

The current high-tech approach risks the same homogeneity that the old Bauhaus produced. This trend contributed to the disappearance of local building cultures and knowledge of harmonious building practices with local ecosystems. Future architects and engineers must not disregard the symbiotic relationship between people, land, and nature. The opportunities posed by climate change adaptation should be utilized to recreate a truly sustainable building culture. The answer lies within vernacular knowledge, which encompasses the architectural language of the people, shaped by ethnic, regional, and local dialects [16]. Contemporary sustainable design often leans on a technocentric approach, presupposing that human societies require improvement through Global North (Western) science. While this approach provides methodologies for quantification and monitoring, it often seeks to dominate and interrogate nature. In contrast, vernacular knowledge sustains reciprocal relationships among nature, culture, and humans. Indigenous communities, through long-term observation, create rigorous scientific approaches known as native science [17, 18]. Unfortunately, vernacular knowledge is often framed as primitive and not advanced or emerging, hindering its evolution into climate change solutions [19]. The profound difference between Global North worldviews and indigenous knowledge systems lies in their relationship with the land. Buildings and cities have historically been tools of colonialism, while vernacular knowledge treats land as part of life and community [20]. Indigenous science interprets nature within the context of 'their place', and vernacular architecture is a body of knowledge built up over generations by culturally distinct people living in close contact with land and its resources [21]. The tipi example in Chapter 8 is a great example.

9.4.3 Development integrating local and vernacular knowledge from the Global South

Here, the author identifies three key areas of sustainable knowledge that can be learned from vernacular and traditional architecture in the Global South: (1) an effective embodied policy must consider contextual factors, (2) embodied carbon reduction should emphasize low-tech solutions, and (3) culturally appropriate solutions must be chosen.

First, an effective embodied policy requires plural but contextual approaches to global challenges. With 194 countries and the EU joining the Paris

Agreement, addressing climate change is a global commitment through political will. Buildings must be constructed to be carbon neutral by limiting both operational carbon and embodied carbon. The Global North's solution to this shift has been mass timber construction, leveraging the renewability and atmospheric carbon sequestration of wood products like cross-laminated and glue-laminated timber. However, the sustainability of this approach is uncertain, especially given the varying forest resources and population growth in different regions. In fact, the largest forest areas allowing for the sustainable sourcing of wood building materials exist in developed countries, notably in North America and Europe [22], while the largest building stock growth happens in developing countries—Asia and Africa—some of which are experiencing significant population growth [23, 24]. Accordingly, the application of timber buildings as a one-size-fits-all approach to all building types may not be a solution for many developing regions.

Research, such as that carried out by Onyefiri et al. and Kubba, showed that sustainable forest management may not be effective everywhere and can even increase deforestation in some low-income economies due to higher foreign investment and international timber demand. In these cases, locally available low-carbon materials, such as straw bale, which has superior thermal properties and is a renewable by-product, can be used effectively [25, 26]. Recent scientific evidence indicates it has superior thermal properties compared to concrete, brick, and wood [27]. Using natural materials in construction based on their local availability can reduce embodied carbon and emission particulates released from burning [28].

Second, design should favour low-tech, low-impact over high-tech, high-impact ones. Ecological economist Herman Daly defined sustainable development as qualitative improvement without a quantitative increase. The development must occur within the biosphere's limits, considering its rebound effect, especially in growing economies where technological advancement can lead to increased resource consumption. Designing for a longer lifespan, spatial and structural reusability, and adaptability can enhance architectural quality and reduce the demand for new construction. This approach contrasts with Global North modern practices that often result in short-lived buildings designed for specific purposes and difficult to disassemble.

Third, design must be culturally grounded. Cultural relevance has been central to architecture for centuries, evolving through trial and error to adapt to communities' living cultures and collective wisdom. An example is *mashrabiya*, a perforated screen made out of cut stone or wood [27] and used as a distinctive feature of traditional architecture in some Islamic countries.

Its design is based on environmental, social, and cultural factors [29], and the screen's porosity serves multiple functions: it enables air circulation, generates shading and protection from intense sunlight, and creates privacy while allowing for outdoor visibility (Figure 9.2). Another example is *Yakhchaal*—ancient Persian ice houses—relied on for their dome-shaped forms, underground structure, and the thermal mass provided by thick adobe to provide a local architectural solution to a practical and climatic challenge (i.e. the need to store ice for summertime use under hot and arid conditions) Figure 9.2. Until the collapse of cultural frontiers in the twentieth century, these distinctive local forms dominated architecture in different societies, and their aesthetics were the co-product of technological availability, environmental responsiveness, practicality, and cultural relevance. In this sense, judging the quality of architecture based on beauty standards from the Global North lens underestimates the complexity of variations in personal preferences and experiences as shaped by factors. Consequently, the concept of singular beauty criteria can lead to a reduction in diversity, plurality, and cultural suitability. The question yet to be answered by the NEB initiative is whose beauty standards it uses as a point of reference.

While interest in vernacular knowledge spans various disciplines, its application in sustainable building practices remains limited. Recognizing the complex, contextual, and dynamic nature of sustainable solutions is crucial,

Figure 9.2 (left) *Mashrabiya at Al-Sadat* house in Cairo. Reproduced from Dr Eomr via Wikimedia Commons (2016) CC BY-SA 4.0. (Right) Yakhchaal, ice house, Iran. Reproduced from Bockomet via Wikimedia Commons (2016) CC BY-SA 4.0.

as opposed to relying solely on technology-driven and optimization-oriented approaches from the Global North.

9.5 Conclusion

As this book's last chapter, Chapter 9 brings to light the intricate relationship between embodied carbon emissions and global inequity, emphasizing the need for a comprehensive approach to sustainability. The chapter integrates multidisciplinary perspectives, revealing that effective solutions must address not only the technological and environmental aspects but also the socioeconomic and cultural dimensions of the issue. The discussions highlight the disproportionate burden borne by the Global South, which suffers the most from environmental degradation while contributing the least to global emissions. This inequity underscores the necessity of rethinking global policies and frameworks to ensure that they are inclusive and equitable. By incorporating local and vernacular knowledge, the chapter advocates for low-tech, sustainable practices that are often overshadowed by high-tech solutions favoured in the Global North.

The proposed action framework outlines strategic priorities, increased R&D in environmental science, avoidance of high-tech biases, and improvements in the construction supply chain. These principles aim to create an equitable approach to reducing embodied carbon emissions. The chapter emphasizes that addressing embodied carbon is not just an environmental issue but a matter of social justice, requiring policies that integrate sustainability with economic strategies and ensure fair carbon pricing mechanisms. Moving forward, the future of global climate policy must prioritize the needs of the most vulnerable populations, fostering global cooperation and leveraging diverse perspectives to create a sustainable and just world. By adopting these comprehensive and inclusive strategies, significant and lasting progress can be achieved in reducing embodied carbon emissions.

References

1. Hu M. A building life-cycle embodied performance index-the relationship between embodied energy, embodied carbon and environmental impact. Energies. 2020 Apr; 13(8).
2. Roberts JT, Parks B. A climate of injustice: global inequality, north-south politics, and climate policy. Cambridge, MA: MIT Press; 2006.
3. Helm D. Net zero: how we stop causing climate change. London: HarperCollins UK; 2020.

4. Helm D. Legacy: how to build the sustainable economy [Internet]. 1st ed. Cambridge, UK: Cambridge University Press; 2023 [cited 6 May 2024]. Available from: https://www.cambridge.org/core/product/identifier/9781009449212/type/book
5. Fenech ME. Maximising the opportunities of artificial intelligence for people living with cancer. Clinical Oncology. 2020 Feb; 32(2): e80–5.
6. Busch J, Lubowski RN, Godoy F, Steininger M, Yusuf AA, Austin K, et al. Structuring economic incentives to reduce emissions from deforestation within Indonesia. Proc Natl Acad Sci USA. 2012 Jan 24; 109(4): 1062–7.
7. Luo J, Yang J, Wan S, Zhao J, Liang J. Can energy conservation and emission reduction policies affect household carbon emissions? Evidence from China. Front Energy Res. 2023 Feb 9; 11: 1138467.
8. OECD. General goverment expenditure [Internet]. [Accessed 5 Jun 2024]. Available from https://www.oecd-ilibrary.org/content/data/f743d6a9-en
9. Abergel, T, Dean B, Dulac, J, Hamilton, I. 2018 Global status report: towards a zero-emission, efficient, and resilient buildings and construction sector [Internet]. 2018 [Accessed 9 May 2022]. Available from https://www.worldgbc.org/sites/default/files/2018%20GlobalABC%20Global%20Status%20Report.pdf
10. UN Environment Programme. 2023 global status report for buildings and construction [Internet]. 2024 Mar [Accessed 9 Apr 2024]. Available from https://www.unep.org/resources/report/global-status-report-buildings-and-construction
11. Enerdata. GHG emissions in developing countries—issues and perspectives for COP-26 [Internet]. 2021 [Accessed 2024 Jun 3]. Available from https://www.enerdata.net/publications/executive-briefing/ghg-emissions-trends-developing-countries-cop26.html
12. statista. Distribution of crude steel production in India in 2022, by process route [Internet]. [Accessed 6 Jun 2024]. Available from https://www.statista.com/statistics/714398/india-crude-steel-production-share-by-process-route/#:~:text=Basic%20oxygen%20furnace%20(BOF)%20was,the%20world%20just%20after%20China.
13. Jingna K. Analysis: China's steel industry needs time to transition [Internet]. Dialogue Earth. 2024 [Accessed 6 Jun 2024]. Available from https://dialogue.earth/en/business/analysis-chinas-steel-industry-needs-time-to-transition/#:~:text=Long%2Dprocess%20steelmaking%20predominates%20in,steel%20is%20made%20this%20way.
14. Naujokaitytė G. MEPs not convinced Europe needs a new Bauhaus movement [Internet]. Science/Business. 2020 [Accessed 2022 Nov 26]. Available from https://sciencebusiness.net/news/meps-not-convinced-europe-needs-new-bauhaus-movement
15. Bergdoll B, Dickerman L. Bauhaus 1919-1933: Workshops for modernity. New York: The Museum of Modern Art; 2009.
16. Oliver P. Built to meet needs: cultural issues in vernacular architecture [Internet]. London: Routledge; 2007 [Accessed 23 Nov 2022]. Available from https://www.taylorfrancis.com/books/9781136424052
17. Johnson JT, Howitt R, Cajete G, Berkes F, Louis RP, Kliskey A. Weaving indigenous and sustainability sciences to diversify our methods. Sustainability Science. 2016; 11: 1–11.

18. Brayboy BMJ, Castagno AE. Indigenous knowledges and native science as partners: a rejoinder. Cultural Studies of Science Education. 2008; 3: 787–91.
19. Maclean K, Greenaway A, Grünbühel C. Developing methods of knowledge co-production across varying contexts to shape sustainability science theory and practice. Sustain Sci. 2022 Mar; 17(2): 325–32.
20. King AD. Colonial urban development: culture, social power and environment [Internet]. London: Routledge; 2012 [Accessed 26 Aug 2024]. Available from https://www.taylorfrancis.com/books/9781135681081
21. Cajete GA. Indigenous science, climate change, and indigenous community building: a framework of foundational perspectives for indigenous community resilience and revitalization. Sustainability. 2020 Nov 17; 12(22): 9569.
22. Kraxner F, Schepaschenko D, Fuss S, Lunnan A, Kindermann G, Aoki K, et al. Mapping certified forests for sustainable management—a global tool for information improvement through participatory and collaborative mapping. Forest Policy and Economics. 2017 Oct; 83: 10–8.
23. Göswein V, Arehart J, Pittau F, Pomponi F, Lamb S, Zea Escamilla E, et al. Wood in buildings: the right answer to the wrong question. IOP Conf Ser: Earth Environ Sci. 2022 Sep 1; 1078(1): 012067.
24. Hamiltton I, Rapf O, Kockat D, Zuhaib D. 2021 global status report for buildings and construction [Internet]. United Nations Environmental Programme; 2021 Oct. Available from https://www.unep.org/resources/report/2021-global-status-report-buildings-and-construction
25. Onyegiri I, Ugochukwu IB. Traditional building materials as a sustainable resource and material for low cost housing in Nigeria: advantages, challenges and the way forward. International Journal of Research in Chemical, Metallurgical and Civil Engineering. 2016; 3(2): 247–52.
26. Kubba S. Green Building Materials and Products. In: Handbook of Green Building Design and Construction, 2nd ed. Oxford, UK: Butterworth-Heinemann, 2017, pp. 257–351.
27. Marques B, Tadeu A, Almeida J, António J, de Brito J. Characterisation of sustainable building walls made from rice straw bales. Journal of Building Engineering. 2020 Mar; 28: 101041.
28. Alcorn A, Donn M. Life cycle potential of strawbale and timber for carbon sequestration in house construction. In 2nd international conference on sustainable construction materials and technologies; 2010, pp. 28–30.
29. Bagasi AA, Calautit JK. Experimental field study of the integration of passive and evaporative cooling techniques with Mashrabiya in hot climates. Energy and Buildings. 2020 Oct; 225: 110325.

Index

(Tables, figures, and boxes are indicated by an italic t, f, or b.)

For the benefit of digital users, indexed terms that span two pages (e.g., 52–53) may, on occasion, appear on only one of those pages.